KINSHIP: BELONGING IN A WORLD OF RELATIONS
VOLUME 3: PARTNERS

VOL. 03

PARTNERS

Edited by

Gavin Van Horn, Robin Wall Kimmerer, John Hausdoerffer

Center for Humans and Nature Press

Center for Humans and Nature Press, Libertyville 60030

For more information, contact the Center for Humans and Nature Press, 17660 West Casey Road, Libertyville, Illinois 60048.
Printed in the United States of America.

Cover and slipcase design: LimeRed, https://limered.io

ISBN-13: 978-1-7368625-0-6 (paper)
ISBN-13: 978-1-7368625-1-3 (paper)
ISBN-13: 978-1-7368625-2-0 (paper)
ISBN-13: 978-1-7368625-3-7 (paper)
ISBN-13: 978-1-7368625-4-4 (paper)
ISBN-13: 978-1-7368625-5-1 (set/paper)

Names: Van Horn, Gavin, editor | Kimmerer, Robin Wall, editor | Hausdoerffer, John, editor
Title: Kinship: belonging in a world of relations, vol. 3, partners / edited by Gavin Van Horn, Robin Wall Kimmerer, and John Hausdoerffer

Description: First edition. | Libertyville, IL: Center for Humans and Nature Press, 2021 | Identifiers: LCNN 2021909501 | ISBN 9781736862520 (paper)

Copyright and permission acknowledgments appear on page 129.

Center for Humans and Nature Press
17660 West Casey Road, Libertyville, Illinois 60048

www.humansandnature.org

Printed by Graphic Arts Studio, Inc. on Rolland Opaque paper. This paper contains 30% post-consumer fiber, is manufactured using renewable energy - Biogas and is elemental chlorine free. It is Forest Stewardship Council® and Rainforest Alliance certified.

CONTENTS

KINNING: INTRODUCING THE KINSHIP SERIES

Gavin Van Horn

The lines hung lightly suspended in midair. Twinkling, illuminated from within. Then vanished. I inclined my head. The lines reappeared, seeming to materialize out of emptiness. These weren't merely lines; they were radial strands precisely strung from a central axis, intricately woven. The glow from a nearby streetlamp caught within them, revealing a sacred geometry. I leaned closer. With a slight tilt of my face—altering the angle between my eyes and the spider's evening project—the lines alternately disappeared or disclosed themselves. Their creator, deftly putting the finishing touches on this work of body art, was smaller than my thumbnail. Her handiwork sparkled with its own radiance. For a moment I felt envious—then grateful. The craft on display demonstrated skills of which I was completely and utterly incapable. I drew in closer to get a better look at the stitching, which would likely hang for the evening before being pulled apart by a strong morning wind.

Kinship: Belonging in a World of Relations can be described as a series of books, five volumes that group different essays and poems according to the scale of their subject matter, from the composition of the cosmos to the gestures of the everyday: Planet, Place, Partners, Persons, Practice. But these books could just as easily be described as a web, a meshwork whose strands gather, crisscross, and link together a vast variety of subjects and experiences. Each book reaches beyond its pages, spinning silk filaments through the others; turn your head at the right angle and an intricate web appears—functional, sensorial, and artful.

The essays and poems you hold in your hands comprise lines, strands of ink, patterns on paper. In your imagination, these words may come to life, recalling and revealing shared relations with our fellow Earthlings—our kinfolk—who come in all shapes and sizes, from the bacterium swimming in your belly or lying on the tip of your tongue to the vibrant collective breath that sweeps across your face and into your lungs. Worth thinking about—and perhaps *thanking* about—are the shared threads between kinfolk, especially plantfolk, that make this breath exchange possible. Your life, my life, all of our lives depend on the quality of relations between us—the air we breathe, the water we drink, the food we eat and the food we become—within an exuberant, life-generating planetary tangle capable of nurturing intelligences that can spin webs and words.

Kinning

The words in this *Kinship* web gesture beautifully toward the relations—vital, wild processes—that are always present yet not always visible. Because these relations may be difficult to apprehend, it may seem as though the world is merely a collection of inert objects, full of nouns. You are you. I am me. That bird at the feeder is, sadly, referred to as "it." That river underneath the bridge and that mountain on the horizon are designated "natural resources." Some of us have rights, legal standing, personhood. Some of us—depending on which nation-state we happen to dwell in—don't.

Nouns have utility, yet they can mislead, perhaps even reifying the idea that the world is composed of things—some small, some large, some shiny, some dull, some with wings, some with legs, some with leaves, some with fur. This language-induced reduction would suggest everything is mere matter, a gathering of atoms, in more or less complicated geometries. Note, however, even in that last sentence, an interesting gerund slipped into the mix. What are atoms but a gathering of *relations*? What is a gathering of relations if not *relatives relating*? Just as when I tilted my head and the spiderweb "appeared," due to a slight change of perspective, it is possible with a shift of perspective to see the threads connecting worlds, all the relations that make us kin. The point here is one that comes up repeatedly in *Kinship*: Earth—and everything within it, including all that creates what we call earth—is a verb. All is in motion; all is relating.

The English language is noun dominant, and in comparison to many Indigenous languages, the animacy and agency of other beings and processes often receive less emphasis. Though obviously still relying on English, these *Kinship* volumes—because of their subject matter—challenge this object obsession at its core. *Kinship* can be considered a noun, of course, a state of being—whether this is couched in terms of biological genetics; family, clan, or species affiliation; shared and storied relations and memories that inhere in people and places; or more metaphorical imaginings that unite us to faith traditions, cultures, countries, or the planet. But the voices in these volumes point us toward an alternative perspective: kinship *as a verb*.

Perhaps this kinship-in-action should be called kinning. Humans are born kin, in any number of ways. But the words in this *Kinship* anthology collectively express something more than birthright claims: they point toward how it is possible to *become* kin. In this understanding, being kin is not so much a given as it is an intentional process. Kinning does not depend upon genetic codes. Rather, it is cultivated by humans, as one expression of life among many, many, many others, and it revolves around an ethical question: how to rightly relate? We are kinning as we (re)connect our bodies, minds, and spirits within a world that is not merely a collection of objects but "a communion of subjects," as Thomas Berry put it.[1] The essays and poetry in these volumes, at different scales and in different geographies, show possibilities for becoming better kin—more receptive to the languages of others, especially nonhuman others, and better listeners to their stories, which reach out to us through place and time. This vibrant world, as well as these volumes, offers invitations for kinning—practices of belonging with and amid our fellow earthly kin.

Three Threads in the Web

Three conceptual threads came together to inform and inspire the creation of the multiple *Kinship* volumes. I'd like to briefly mention them here, as readers may want to be alert to them throughout various essays in the books.[2]

The first thread is a cosmovision—and, increasingly, legal recognition—that acknowledges and understands nonhumans, including entire watersheds, forests, or mountains, as persons. In the West, many of us have inherited a settler-colonial worldview that uplifts the human individual—often implicitly or explicitly identified with whiteness and maleness—as the locus of meaning and center of importance while reducing nature to resources, property, or fungible commodities. Bootstrap economics and the literary hero's journey reinforce such thinking—the lone figure encountering and overcoming obstacles, conquering beasts, and emerging victorious above the fracas. From this vantage, there are human persons (and now corporate "persons"), and there is everything else.

The religious studies scholar Graham Harvey, in his wide-ranging study of animistic and neo-animistic cultures and movements, upends such notions. Harvey observes that from an animistic perspective, "the world is full of persons, only some of whom are human." Persons, he goes on to write, are not equated solely with human beings in many cultures. Rather, the term serves as a broader umbrella for those beings who are perceived as displaying agency (and this encompasses landscapes, rocks, and bodies of water, in addition to plants and nonhuman animals):

> Persons are beings, rather than objects, who are animated and social towards others (even if they are not always sociable).

> Animism *may* involve learning how to recognize who is a person and what is not—because it is not always obvious and not all animists agree that everything that exists is alive or personal. However, animism is more accurately understood as being concerned with learning how to be a good person in respectful relationships with other persons.[3]

When I first read Harvey fifteen years ago or so, I didn't know how or if this type of cosmology could ever make its way into mainstream Western consciousness. Then, in March 2017, the Whanganui River (Te Awa Tupua), the third-largest river in Aotearoa New Zealand, grabbed international headlines. The Whanganui officially gained legal status as a living entity with the same rights of personhood as a human being. More than a change in legal nomenclature, this reclassification of the river stands as a significant bicultural effort to bring disparate systems of law and care together among New Zealanders of European descent and the native Māori population (Te Āti Haunui-a-Pāpārangi).[4]

The designation of the Whanganui River is one instance in a growing number of cases in which legal personhood is being granted to nonhuman entities. An overlapping set of localized and national governmental precedents, many of which involve personhood language, for example, began to gain traction in 2006 by focusing on the "rights of nature."[5] Ecuador and Bolivia both included rights-of-nature clauses in their national constitutions in 2008 and 2010, respectively. In Colombia, courts ruled in favor of personhood for the Amazon and Atrato Rivers. In 2016, the Ho-Chunk Nation in Wisconsin amended their tribal constitution to include rights-of-nature language: "Ecosystems, natural communities, and species within the Ho-Chunk Nation territory possess inherent, fundamental, and inalienable rights to naturally exist, flourish, regenerate, and evolve." In 2017, the Ponca Nation in Oklahoma recognized rights of nature as statutory law to combat fracking. Australia, India, and Nepal have also taken steps toward establishing rights of nature. In 2019, the Yurok Tribal Council passed a resolution that declared the personhood of the Klamath River in the Pacific Northwest. Such landmark legal and legislative actions represent efforts to give "voice" to other-than-human beings, ensuring their inherent rights to exist and flourish. Gerard Albert, the lead Maori negotiator on behalf of the Whanganui *iwi* (tribe), summed up this sense of responsibility well: "We can trace our genealogy to the origins of the universe. And therefore, rather than us being masters of the natural world, we are part of it. We want to live like that as our starting point. And that is not an anti-development, or anti-economic use of the river but to begin with the view that it is a living being, and then consider its future from that central belief."[6]

Recognition of kinship has many overlaps with these attempts to recognize personhood. The commonality lies in a respect for the agency of other beings and concerted efforts to treat them with dignity and even deference. This brings us to the second thread. *Kincentric ecology*, a phrase coined by ethnobotanist Enrique Salmón, provides a helpful guide for understanding kinship: an intertwining of the social, mythological, and practical. Salmón asserts that "life in any environment is viable only when humans view their surroundings as kin; that their mutual roles are essential for their survival."[7] This perspective stands in marked contrast to the familiar, if not predominant, human chauvinism toward other species in so many national sociopolitical systems. From a kinship perspective, the landscapes of which humans are a part—including rocks, rivers, oceans, prominent geographic features, and other nonhuman plant and animal persons—provide a shared sense of place and require appropriate human care and respect.

This kinship is deep and wide—and dwells within the human body. In the past century and a half, evolutionary and ecological sciences have brought additional insights to bear on what it means to be human. In only the past few decades, evolutionary models are being transformed by research into symbiotic mergers at the cellular level, horizontal gene transfer, and seemingly chimeric creatures that rely on cooperative relationships between species from entirely different "kingdoms" of life. Kinship, it would seem, is key to understanding biotic and abiotic entanglement. A kincentric ecology emerges from cultures that recognize the importance of humans in maintaining right relations in particular landscapes. Far from presupposing that humans are a degrading force, sullying whatever we might touch, a kincentric ecology expresses the view that humans can actually play keystone roles in our landscapes, creating mutual flourishing. In other words, human beings are not merely kin by biological relation, but it is entirely possible that human communities and cultures can be good kin, salutary ecological collaborators alongside and with our nonhuman family members.

The third thread that inspired this *Kinship* series, I'm happy to say, comes from coeditor Robin Wall Kimmerer. Robin's work draws from scientific training and Indigenous knowledge in complementary ways. In *Braiding Sweetgrass: Indigenous Wisdom, Scientific Knowledge and the Teachings of Plants,* she explores her own history of loss and recovery as a member of the Citizen Potawatomi Nation and describes how Indigenous perspectives can transform engagement with a living world. Perhaps nowhere is this clearer than when she contrasts the "grammar of animacy" embedded in the Potawatomi language to conventional English and its objectifying pronouns. She makes a convincing case that an ethical revolution might depend on a language revolution. Finding ways to properly and respectfully acknowledge *ki* (the pronoun Robin proposes for our other-than-human kin) is a good place to begin.

You will recognize these three threads—nonhuman personhood, humans as relational participants in local ecologies, and the care expressed when addressing and engaging with our kinfolk through language—winding their way through all the volumes of *Kinship.*

Five Scales of Kinning

With all the amazing contributors gathered in this *Kinship* web, it might help the reader to know what we as editors were asking of them. The following are the questions we posed to our contributors for each *Kinship* volume, which feature the ways kinship can be understood at different scales: from deep time cosmic and evolutionary relationships; to community watersheds, landscapes, and bioregions; to interspecies engagements and mythological perceptions; to biological and symbolic understandings of human interbeing; to which kinds of practices are appropriate for making and becoming kin:

> *Volume 1: Planet*—With every breath, every sip of water, every meal, we are reminded that our lives are inseparable from the life of the world—and the cosmos—in ways both material and spiritual. What are the sources of our deepest evolutionary and planetary connections, and of our profound longing for kinship?
>
> *Volume 2: Place*—Given the place-based circumstances of human evolution and culture, global consciousness may be too broad a scale of care for us. To what extent does crafting a deeper connection with Earth's bioregions reinvigorate a sense of kinship with the place-based beings, systems, and communities that mutually shape one another?
>
> *Volume 3: Partners*—How do cultural traditions, narratives, and mythologies shape the ways we relate, or not, to other beings as kin? How do relations between and among different species foster a sense of responsibility and belonging in us?
>
> *Volume 4: Persons*—Kinship spans the cosmos, but it is perhaps most life changing when experienced directly and personally. Which experiences expand our understanding of being human in relation to other-than-human beings? How can

we respectfully engage a world full of human and nonhuman persons?

Volume 5: Practice—From the perspective of kinship as a recognition of nonhuman personhood, of kincentric ethics, and of *kinship* as a verb involving active and ongoing participation, how are we to live? What are the practical, everyday, and lifelong ways we *become* kin?

We invited our contributors because of their experiences, their expertise, their diverse backgrounds and geographical locations, and because of the way they've made kin with particular species—or some combination of all of these. We also invited our contributors to share their words because of their abilities to tell a good story.

In many of the essays in *Kinship*, there are statistics, references to academic sources, explorations of complicated ideas, and endnotes that may take a reader onward, but we above all wanted readers to hear people's stories. As human beings, we are storytelling animals. We lean a little closer when we overhear someone else say, "Oh my, have I got a good story for you!" For similar reasons, our Paleolithic ancestors likely leaned into the firelight—as it crackled against the cave walls at Lascaux, France, or Sulawesi, Indonesia—watching the aurochs, bison, horses, and deer dance before their eyes, or the warty pig and the babirusa (pig-deer).[8]

As storytelling creatures, when thinking of our personal relationships with the natural world, we may be predisposed to be on the lookout for epiphanies—the holy overwhelm, the big payoff, the road-to-Damascus moment, the final boss battle. But becoming kin, as the various stories in these volumes attest, consists of repeated intimacies, familiar encounters, and daily undoings and transformations that are dependent on visitations and conversations within a smaller circle of place. Awe-filled moments of raw contact with forces that relativize human importance should not be disparaged or discounted. Such experiences may stir deep wells of gratitude. But knowing that humans are relatives, responsible for the lives of others, as others are responsible for ours, remains at the heart of these volumes. If humans are relatives relating, not merely in terms of an abstract genetic code but as intimate familiars, the question then becomes how we as individual persons and communities might better cultivate these relations. How can we uproot the desire to impose our will upon the living worlds around us? How do we become more receptive to nonhuman languages and ways of being?

One step in this direction is the recognition that nature is not a passive object, a text awaiting our interpretation or exegesis, a thing humans approach solely for insights, entertainment, and "resources." The world all of us are part of and participate in is a relational exchange—alive, wildly generative, an ongoing conversation of bodies, desires, conflicts, and collaborations. There is no pinnacle here for humans to sit atop and gaze upon the masses. *Kinship* culminates, in volume 5, with a wide-ranging conversation about practices and ethics that embrace a world of other-than-human persons as worthy of our active care, concern, and respect. At a time when human fidelity to the natural world seems to be fraying, *Kinship* offers stories of solidarity, highlighting the deep interdependence that exists between humans and the more-than-human world. It explores challenging questions, including how communities might fairly and effectively give voice to nonhuman beings and landscapes. And it highlights the cosmologies, mythical narratives, and everyday practices that embrace a world of other-than-human persons as worthy of response and responsibility.

Humans will survive and continue to tell our stories if we learn how to live well with our kin. The voices included in these volumes—these webs of words—and the collective wisdom they express, invite us into this kind of kinning. These are the stories of how to listen to voices other than our own.

Lean a little closer into the firelight. Up on the ceiling of the cave, or near the streetlamp, or between branches in the forest, or in the corner of the room within which you are currently reading, a web may be flashing in a flicker of light.

NOTES

1. Thomas Berry, *Evening Thoughts: Reflecting on Earth as Sacred Community* (San Francisco: Sierra Club, 2006), 149. More on Berry as "geologian" can be found at the website of the Thomas Berry Foundation, http://thomasberry.org/life-and-thought/about-thomas-berry/geologian.
2. In addition to the themes, the three persons mentioned all have essays in the *Kinship* volumes. Graham Harvey, "Academics Are Kin, Too: Transformative Conversations in the Animate World," appears in *Vol. 4: Persons*; Enrique Salmón, "A Heart Rooted in Place: Poetic Dentists and Getting Rained On," is in *Vol. 2: Place*; and Robin Wall Kimmerer, "A Family Reunion near the End of the World," is in *Vol. 1*: Planet.
3. Graham Harvey, *Animism: Respecting the Living World* (New York: Columbia University Press, 2006), xi.
4. See Anna M. Gade, "Managing the Rights of Nature for Te Awa Tupua," *Edge Effects*, September 5, 2019, https://edgeeffects.net/te-awa-tupua/. As my coeditor Robin pointed out to me, "This can be understood, not [that] the river *gained* personhood" but "more that Western institutions came to acknowledge its intrinsic personhood, under the tutelage of the Maori, who have always recognized this inherent nature."
5. Craig M. Kauffman and Pamela L. Martin, "Constructing Rights of Nature Norms in the U.S., Ecuador, and New Zealand," *Global Environmental Politics* 18, no. 4 (2018): 43–62, https://www.mitpressjournals.org/doi/pdf/10.1162/glep_a_00481.
6. E. A. Roy, "New Zealand River Granted Same Legal Rights as Human Being," *The Guardian*, March 16, 2017, https://www.theguardian.com/world/2017/mar/16/new-zealand-river-granted-same-legal-rights-as-human-being.
7. Enrique Salmón, "Kincentric Ecology: Indigenous Perceptions of the Human-Nature Relationship," *Ecological Applications* 10, no. 5 (2000): 1327–32, https://www.researchgate.net/profile/Enrique_Salmon/publication/242186767_Kincentric_Ecology_Indigenous_Perceptions_of_the_HumanNature_Relationship/links/5c34e542a6fdccd6b59c2aa1/Kincentric-Ecology-Indigenous-Perceptions-of-the-HumanNature-Relationship.pdf.
8. Leang Timpuseng is the name of the cave in Sulawesi, Indonesia, whose figurative art has been geochemically dated using recently developed techniques. The results pushed back the timeline on some of the earliest known figurative cave art, and examples of symbolic thinking, to more than 35,000 years ago. "Find early paintings, particularly figurative representations like animals, and you've found evidence for the modern human mind," writes Jo Marchant in "A Journey to the Oldest Cave Paintings in the World," *Smithsonian Magazine*, January–February 2016, https://www.smithsonianmag.com/history/journey-oldest-cave-paintings-world-180957685/. The timeline continues to lengthen; in 2021, from a karst system in Sulawesi, a date of 45,500 years ago was announced for the earliest known representational work of art. See Adam Brumm et al., "Oldest Cave Art Found in Sulawesi," *Science Advances* 7, no. 3 (January 13, 2021): https://advances.sciencemag.org/content/7/3/eabd4648.

A PRAYER TO TALK TO ANIMALS

Nickole Brown

Lord, I ain't asking to be the Beastmaster
gym-ripped in a jungle loincloth
or a Doctor Dolittle or even the expensive vet
down the street, that stethoscoped redhead,
her diamond ring big as a Cracker Jack toy.
All I want is for you to help me flip
off this lightbox and its scroll of dread, to rip
a tiny tear between this world and that, a slit
in the veil, Lord, one of those old-fashioned peeping
keyholes through which I can press my dumb
lips and speak. If you will, Lord, make me the teeth
hot in the mouth of a raccoon scraping
the junk I scraped from last night's plates,
make me the blue eye of that young crow cocked to
me—too selfish to even look up from the flash
of my damn phone. Oh, forgive me, Lord,
how human I've become, busy clicking

what I like, busy pushing
my cuticles back and back to expose
all ten pale, useless moons. Would you let me
tell your creatures how sorry
I am, let them know exactly
what we've done? Am I not an animal
too? If so, Lord, make me one again.
Give me back my dirty claws and blood-warm
horns, braid back those long-
frayed strands of every nerve tingling
with all I thought I had to do today.
Fork my tongue, Lord. There is a sorrow on the air
I taste but cannot name. I want to open
my mouth and know the exact
flavor of what's to come, I want to open
my mouth and sound a language
that calls all language home.

THESE WILD, VIBRANT, UNSTOPPABLE EXPRESSIONS OF ALIVENESS

Martin Lee Mueller

the end is,
grace—ease—

healing,
not saving.

singing
the proof

the proof of the power within.
—GARY SNYDER

Lean into me.
The universe
sings in quiet meditation.
—SIMON ORTIZ

Krrrrrk! Dandelion yearns so hard to stretch toward the sun, it cracks right through pavement. Green stalk, unwounded, hungers skyward. Sun petals thirst open in a perfect circle of yellow. One day the flower closes, dreams some unspeakable sorcery, and reopens into a perfect white globe of feather-winged seeds. Wind-breath caresses the globe; dandelion quivers and lets go. Her many possible futures sail off into the fragrant midsummer ocean of air.

Aliveness blossoms from within the many finned, webbed, barked, or naked bodies we are.

Arctic tern cracks open her eggshell from within. Blinks, chirps, listens. Waits. Sticky wet fluff gripping her heaving rib cage. One day, weeks later only, she stretches her wings, takes to the air, and cruises the planet from pole to pole. In journeying, she assumes her place among the ranks of summer chasers, sun chasers. No other animal on Earth will see as much sunlight as her. A fistful of fiery determination, able to migrate ninety thousand kilometers every year for decades to come. If need be, sun-bird will sleep bedded on an updraft. She is born a sailor. Farthest of them all. Place connector, boundary defier (inner and outer). A seamstress who weaves the fabric of the breathing whole a little tighter together with every crisscrossing dreamline.

The whole blooms into being through every living body's meaningful forms, gestures, poetic expressions.

Three sea wolves paddle across a stretch of ocean eight kilometers across. A huge black body breaches beside them, blasts out a breath-fountain, vanishes again. The wolves clamber from the ocean amid piles of bleached driftwood. Dripping moss beards drool from the lower branches of thousand-year-old cedars. Three shadows swish through cold mist. Now they freeze! There: rancid, stinking flesh-mountain! Half-liquefied, it has summoned these most oceanic of the *Canis* lineage to this remote outpost of the Great Bear Rainforest. Two dozen ravens and a lone bald eagle protest the sea wolves' arrival. Hisses, whimpering, a yelp. *Caw-caw, caw-caw.* Black wings flap. A muzzle snaps. Integral formalities of a community feast enacted, and reenacted, for generations uncounted. She wolf sinks her teeth into foul flesh. Crunches, chomps. Flies buzz. Maggots doze. Feathers swish. All have come to honor the dead humpback's final sacrifice.

Kinship is being enacted in cascading gestures all around and within.

A thousand Chinook salmon, or perhaps ten thousand. They know no adequate embodied response to the dam that blocks the Elwha River. This. It cannot be. It has never been. Piped water thunders. Ten thousand incarnate river intelligences, dumbfounded. Ten thousand cold-blooded testimonials that water—circulating as willful body, circulating with intention across ancient migration routes, circulating children's stories and a grizzly's winter fat—can leap even against gravity. They have never done anything but leap against obstacles. They leap against the dam. They do not stop. Stopping is no viable body-thought inside their sizzling flesh. They crush their heads trying. They color the river red with their blood. A full century passes. Every year they return. Every year they leap. Every year they crush their heads. Every year they bleed to death. The urge to leap is so powerful a thought, stopping remains simply unthinkable.

Life is a self-birthing habit of coordinated interactions performed on the nonnegotiable premise that life shall be.

She has spent a lifetime loving forth aliveness through lobbying, campaigning, resisting, lecturing, petitioning. She will not stop in this lifetime. Jail time never stopped her. Logging trucks never stopped her. Bullies never stopped her. From her cabin window, she sees sea otters floating on their backs, cracking sea urchins on rocks they've placed onto their bellies. She places her two palms against the window. The pane shudders with the boom of another tourist waterplane as it takes off from the bay. Sometimes she cannot take the hurt. Sometimes it all gets too much. She knows those times will come. She is prepared. She vanishes into the rain forest behind her cabin. The mud trail and her feet have danced this feverish dance forever. She rushes to the spot, she grabs the baseball bat, she raises it above her head, just in time, for she cannot hold back the scream any longer as the bat comes swinging down *hard*! The stained Styrofoam mattress receives blow after anguished blow. The forest receives scream after anguished scream.

Life ever yearns to unfold into deeper relationship.

"What do Atlantic salmon mean to you?" I ask. I wait. "Seven thousand years," he says at last. Three words. They demand patience. To ooze into the space between us. To breathe heat into the burning logs at the center of the teepee. To mingle with the wind in the rustling birches outside. To trickle into the river that grinds past Čearretsuolu island on all sides, as it surges to the Barents Sea. "That is how long my family has lived on these banks. We Sámi were here long before nation-states declared the Tana River to be a border between two separate countries. We were here long before they curbed our fishing rights, sent us to boarding schools, denied our languages. We will not be shamed any longer. We will not have our fishing rights curbed any more. We will not leave this island between Norway and Finland until we are being heard." His voice, just emerging into manhood, speaking with an authority bequeathed by kinship ties reaching into time immemorial. Words as resolute, serene, and vast as the tundra all around us. The words grind past me on all sides. A mosquito stings my hand. I do not withdraw but sit still. I take a deep breath of birch-smoke midsummer air.

What is it to try to honor kinship in a wounded world, in a wounded time?

The animal rights lawyer cries silent tears at night by her kitchen table, a half-empty glass of water in front of her. Words can be weapons. Silence can be a weapon. Yet another complaint letter to the government rejected. Norway will go ahead and shoot a young family of wolves inside the designated predator-protection zone. Make the kill efficient. A ten-minute helicopter chase supersedes fifty hunters pursuing the wolves through rugged terrain for days. It is in the animals' best interest, authorities say. Beside her is a note, an unpaid electricity bill. A life in the economic margins, in the world's richest country. Familiar crack in human-human kinship ties, crack of a hegemonic story that rewards not acts of aliveness but acts of destruction, not deeds of love but deeds of ignorance.

Heartbreak. Alienation. Confusion. Anger. Despair. Loneliness. Ecocide felt from within living bodies. Festering wounds inside the flesh of the real.

Words can be weapons. And silence can pierce deeper than a scream. He lost his voice the day they handcuffed him and carried him off his land. He had refused to leave. For 280 days and nights he had stayed on Swanson Island, his people's ancestral land. Unceded territory. He said he would not leave until the Norwegian salmon-feedlot corporation had left their waters. Chinook salmon, their staple, needed to be given a chance to rebound from near extinction. A Canadian judge ruled otherwise. After the law had spoken, not a word from him. He could not.

Every sacrifice made in honor of true aliveness is a pledge of allegiance with all our relations.

"Papa," says the five-year-old, yawning. "Is the world really a thousand years old?" Her quiet breathing fills the darkness. In. Out. In. Out. At long last a sigh, deep from within the child's chest. "Woooow!" Another outbreath, and she falls asleep, landing soft in the bosom of existential wonder.

The universe awakens to richer inwardness in every child of every lineage of each of life's kindoms.

Tahlequah watched her baby die half an hour after she had birthed it. For sixteen months their two bodies had been inseparable. They had flowed back and forth as blood, heartbeat (first one, then two), warmth, voice, marvel, dream. Had she seen this coming? For three years, every mother in her dwindling kinship clan—the Southern Resident community—had birthed stillborns. Chinook salmon, their staple, gone. Rivers of fatty milk, dry. When her calf breathed out one last time, she was determined. Her baby started to sink into the waters of the Salish Sea, all four hundred pounds of it. She would not let that happen. She nudged it back to the surface, where air flows. It started to sink again. She nudged it back up again. So it went. For seventeen days. For one thousand miles. A dead orca baby kept afloat by its starving mother who knew she had nothing left to give but her grief.

Life stubbornly refuses to give in silently.

Is it possible that in acting to become more deeply integral to the world's unquenchable renewal into meaningful forms, we may be fulfilling the universe's deepest wanting? Is it possible that we can verify this, not by recourse to a certain disembodied objectivity, but by turning more attentively toward the poetic objectivity that grows from experiencing aliveness from within living bodies? To be granted the gift to participate in ever-rejuvenating moments of mutuality through interbeing! To commemorate the countless ways in which life yearns to experience itself as intimate presence! To dwell in the unfathomable immensity of Another! To live in the openness of life's inexhaustible fecundity! The nucleated cell, billion-year-old communion of former opposites. The apple flower whose every gesture desires the touch of bumblebee. The dust-covered rooster who daydreams in the blissful abandon of his midday sun bath. The crisscrossing ant trails on the sunbaked pine-needle forest floor, corporeal infrastructures of complex social composition. The sharp, frozen sky that presses itself against the toddler's cheek one winter solstice dawn. The double-winged seed that comes flying from across the lake, not unlike a butterfly, only to settle down in a moist granite crack, where it germinates and begins its lifelong habit of embodying light, embodying water, embodying rock, until decades later, airborne seed has shapeshifted through the touch of all into rockborne birch elder. Who is birch, if not the patterned dance of mineral flow, water flow, air flow, a self-making web of co-ordination, felt from within a wooden body? What is the biosphere, if not a constantly unfolding, actively composed sympoetic sphere of meaning making in mutuality? What is embodied inwardness, if not an immanent quality of a univocal universe? What are all these wild, vibrant, unstoppable expressions of aliveness, if not variations of the one voice of Being? Is not each breathing body an actualization of matter self-composed so complexly that it desires to express itself from within—as feeling, thought, thirst, arousal, dejection, passion, or unquenchable yearning to live?

Such a view of aliveness as immanence would take its cues from Spinoza, first of Western philosophers to articulate an ontology of immanence, and from later philosophies of immanence such as those of Henri Bergson, Thomas Berry, Gilles Deleuze, David Abram, or Andreas Weber. From such a point of view, to ask again who we are as human beings is also to ask again what—or who—else participates in this biosphere's unstoppable outpouring into self-expression. Acts performed in defense of life, from this point of view of immanence, are concrete embodiments of life's yearning to be. They may be politically inopportune. They may be economically sidelining. They may be socially stigmatizing. They may be personally defeating. They may starve our bellies, silence our voices, drain our hearts, break our necks. But they matter. Every one of them matters. Actions that celebrate, honor, or guard life matter as an ontological manifestation of aliveness, a reorientation toward Being as Being-in-participation. Life's upwelling into deeper kinship ties cannot and will not be stopped, even as Earth is now shuddering with extinction spouts. Through this living Earth, the universe has infected itself with a wanting deeper than human affairs. It is a deep-time infection, moving forward through time as a cascading river of ever-unfolding possibility, embodied through living beings who want more of that precious stuff: life. It is a river that will not be forever dammed. Even as many beautiful things die, we can recognize that every gesture toward strengthening aliveness creates potential for new expression, helps realize the existential factualness of kinship, and gifts forward novel opportunity. As such, each such gesture is inalienably valuable. It is also inherently beautiful.

It matters what we do, even when from within a narrow, utilitarian imagination it may seem futile or naïve to leap against a concrete wall in a river, or to leap against a concrete wall in the imagination. No act of honoring aliveness is ever futile or naïve. That's because every act of honoring aliveness already actualizes aliveness concretely. In a world of constantly unfolding kinship relations, no such act is ever separate. It offers itself as thread, step, voice, fertilizer, performance, womb, spark: toward more life to come. As it will.

Sacrifice, etymologically speaking, is the act of making sacred.

And ever the salmon leap. On the Elwha River, a century came and went. Their numbers collapsed. Stopping was never a viable body-thought. Some old-timers said the fish never stopped reminding us of our obligation to honor them. We failed them, but no longer. We are beginning to listen to the world again. Both dams on the river are now dismantled, in what made international headlines as the world's largest dam removal, anywhere, ever. The salmon? Within weeks of the dam removals, they leapt upriver. What else? They instantly got to business, cocreating once again a more vibrant, more intricately entangled existence.

The end is grace—ease—

Tools of destruction, through some unspeakable sorcery, can become tools of healing. It took but a bold leap of the imagination to reimagine logging machinery as reforestation tools. With their help, the Klamath River in Northern California may one day again reach the ocean, year-round. Cycles of degeneration will have been reversed into cycles of regeneration. The temperate redwood rain forest will bring the rains back. Salmon will spawn again throughout the river's arteries. Beavers will again help regulate the water table. Trophic cascades will again ripple out into the soils, the canopies, the hillsides, the rivulets, the streams. The people—Indigenous and of recent arrival—are already crafting new expressions of belonging, new kinship ties. New kindigeneities.

Healing, not saving.

Six thousand liters of chlorine spilled into the Akerselva—my home river in downtown Oslo—during the night. A technical failure in the water treatment facility, by all accounts. The poison sterilized the river to the microbial level. Killing all. Half a dozen years later, salmon are leaping again. They leap up the rapids, up the fish ladder, against the waterfall. Streetcars screech. Traffic booms. Joggers pass. Young fathers push their baby carriages along the banks. And the salmon? With noses pointing upriver, they are nearly still. They've completed their journey. At long last they have returned to their headwaters. A new generation will grow from here. It is time.

Singing the proof.

He found his voice back. Young chief of the Namgis First Nation! He stood before an audience of a hundred, welcoming us as guests to their land. This was the first time he could speak again in public. His newfound voice: eloquent, vibrant, unstoppable. His words painted the image of a future in which, one day, his children and children's children would again be able to cross rivers on the backs of fish and never get their feet wet. That's how many fish there were in the past. That's how many there will be yet again in the future.

Now. This inexhaustible spring of grace.

LISTENING TO A RIVER'S LAW

Ourania Emmanouil

This river runs dark brown. It is stained with tannins, and I struggle to see more than a foot below its surface. My back rests against a eucalypt that is rooted into the riverbank, branches reaching across the water, leaves brushing the flowing stream. The breath of the river touches my face and the whole city seems to exhale.

While this river has known me since childhood, I cannot say that I have really known it. Born of migrant parents, I grew up and have returned to one of her southern watersheds, just south of the Koonung Koonung Creek. A few years ago, a new "face" of the river became visible to me and other settler people living in Melbourne, Australia's second-largest city. A new act of parliament—the Yarra River Protection (Wilip-gin Birrarung murron) Act of 2017 (Victoria) (the Birrarung Act)—recognized the Yarra River as a single, living, integrated natural entity.

This shift away from an instrumental or propertied understanding of the river is significant. It comes after decades of advocacy by people who love the river, shifting relations between the state and First Peoples and inspiration from legal developments in Aotearoa/New Zealand. It signals a movement toward a much older knowing of the river. This new law calls the river by its ancestral name, Birrarung, and invokes the river's kin in Woi-wurrung language. In doing so, it makes apparent a much older law and set of legal relations. The Birrarung Act tells an old story about keeping the Birrarung and its relations alive. It is the first time that settler law on the Australian continent has been written, in part, in an Aboriginal language, and has invoked ancestral beings. The preamble to the Birrarung Act speaks of the lawful relations that hold in balance Birrarung and its kin: Bunjil, the great eagle creator; Waa, the crow who is protector; the Kulin people made from the earth; Palliyang, the bat creator of Bagarook (women) who came from the waters of Birrarung. Outside of the Birrarung Act, knowledge of these connections is held in and passed down through Dreaming stories.

While Birrarung and its kin have always been around me, it was not until reading about this web of legal relations that I became attuned to Birrarung's law. Through this relational understanding of law, the river morphed into something more than just a river in my own consciousness and, perhaps more broadly, in the collective settler imagination.

To "see" and "listen" to Birrarung's law, performed through these kinship relations, is to take seriously other-than-Lockean cosmologies. In doing just this, the Birrarung Act has the potential to generate dialogue across cultures and legal orders and to reconcile divergent understandings of what the river is and how it should best be protected. The Birrarung Act invites people who dwell with and are responsible for the river—both First Nations and settler peoples—to enact relationships of care and respect with Birrarung. This law calls for settler peoples to imagine the river as something more than just a waterway, creating an opening for new distinctions to be drawn around what it means to be lawful. For me, it holds the possibility for exploring how I, as a settler person, can honor and respect First Peoples' laws and cultures in my action to protect the Birrarung and other nonhuman nature.

The lawful relations enacted by Birrarung and its kin raise this question: what might it mean for me to be lawful in and with this place? To ground this in my own experience of kinship, we first need to leave the Birrarung and go to Mäpuru, a homeland in northeastern Arnhem Land, where a patterned understanding of the universe awaited me.

Waking Up to a Universe Patterned in Kinship

Bright-colored strands of pandanus fiber lay across my lap as I sit cross-legged on a woven plastic mat. Four generations of Yolŋu weavers here in the Mäpuru homelands patiently teach me and ten other visiting Balanda (non-Indigenous) women basket weaving through slow demonstration. Starting the coil is always the hardest part of weaving a basket. I have no coordination or rhythm to my movement, until Bambalarra reaches in and gently guides my hands in a way that makes tactile sense. As matriarchs of this community, Bambalarra and her sister, along with their daughters, granddaughters, great-granddaughters, and other kin, ground this practice. It is more than just weaving, it is *gurul'yun,* being with.[1]

In between stitching, I lean back and look at the baskets swaying from the rafters of the weaving shelter and become aware of the possibilities of form, color, and pattern. Delvina, a young woman in the community, invites me to become her *yapa,* or sister. She extends to me her *gurruṯu* (kinship) and *mälk* (skin)—*baŋaḏitjan.* As a recursive pattern that applies to everyone and everything, *gurruṯu* acts to situate me as a relationally bound entity in the Yolŋu universe. It tells me that the hard boundaries between self and other that I have lived by are not so—I am constituted by all these entities and they of me. *Gurruṯu* teaches me how to be respectful, what my responsibilities are, and how to maintain balance in and with the world. Word spreads that I am *baŋaḏitjan* and Delvina's *yapa.* This new relational identity, part of something much bigger, is invoked by this new name. I begin to feel my identity as an atomized individual dissolve at the edges.

Before Mäpuru, I saw myself as part of a constellation—a rambling web of Macedonian ancestors and family, from the same village in northwestern Greece and now mostly spread across the suburbs of Melbourne. While I felt my connections to these people and the Wurundjeri Country that we now call our home, there was something more that had not yet revealed itself to me.

Being patterned through *gurruṯu* was the first step in me recognizing and enacting what Tyson Yunkaporta calls the kinship-mind. Yunkaporta says that "nothing exists outside of a relationship to something else."[2] *Gurruṯu* expanded my constellation. When I closed my eyes, I could see my connections spreading between human and more-than-human kin. A sunset was kin. A particular wind was kin. Yams were kin. Places were kin. More than this, there were words that I could invoke that spoke directly to the reciprocity of these relationships.

Coming home to Wurundjeri Country, I had none of this language, no words or stories to "see" and "know" who or what might be kin. It was years later that the possibility of kinship with my birthplace became apparent.

What Is a River?

When I look back at my evolving relationship with Birrarung, I can see that an interplay of stories, language, and listening broadened my greater recognition of what this entity is and, more pertinently, how it performs. The words of Ellen Van Neerven speak to this relationship: listening "is a first act of learning, and through learning we become not only aware, but appreciative of what is, and always has been, around us."[3] Within my direct, embodied experiences of being with Birrarung, there is a deep listening and acknowledgment. Yet listening to the stories of Birrarung and paying them the respect they warrant brings another layer of awareness to this relationship. By observing how the river performs as cultural, political, and legal actor in stories—new and old—I can critically reframe myself as a legal actor and explore the potential to become kin.

Birrarung as Law

Indigenous legal philosophers Mary Graham and Christine Black offer two fundamental axioms that disrupt Western positivist understandings of what law is and how it lives: land *is* the Law, and it is *lived* through the Law of Relationship. These axioms also upset the assumption that laws originate solely from parliament and the courts. Black describes First Laws as offering a "sacred formula to produce good health in a knowledge system based on being within . . . a productive/creative complimentary system of relationships."[4] The old story carried forward by the Birrarung Act attunes me to the network of legal relations that constitutes Birrarung. It offers me ground to engage in respectful and generative dialogue with Wurundjeri from a place that recognizes and values the ongoing performance of First Laws.

As a settler person living on an ancient continent with hundreds of legal orders, I contemplate how can I be respectful on this law-filled land when there are "limits" and "boundaries" that protect First Laws—how they can be known and by whom. My teachers on Country and deep listening practices, which I refer to as *liyan* (feeling, intuition, connection), sometimes offer me glimpses into the legal patterning in Country.[5] Although I will likely never understand this law, the guidance offers me a foundation to respect the land and Birrarung *as* law.

Birrarung as Country

This ancient river, Birrarung, has nourished all of its people—human and more-than-human—for millennia, holding law, knowledge, stories, and songs. As a place-based, multidimensional entity, "Country" is held together by reciprocity, relationality, and respect. Human kin have a responsibility to maintain and enliven Country, which gives in return. This ethic of reciprocity of care is embedded in the Birrarung Act. While this reality continues to be lived by First Peoples across the Australian continent, historical and present-day colonial conflicts over land reflect a deep, unresolved, and unconscious ontological politics—or contestation over what the land is and how it should be valued.

Birrarung as a Site of Contested Cultural Meaning

Violence (still) irrevocably shapes politics enacted in and with Country. Perpetuated through ongoing processes of neo-colonization and historical displacement and cultural genocide, it has supplanted old identities, renamed what "exists" and attempted to sever ancient associations. What lived as songline, ancestor, and teacher with its Wurundjeri and Boon-wurrung kin took on a very different meaning and life in the imaginations of colonizers and settlers.[6] A colonial blindness and instrumental culture, propagating a propertied view of the world, simplified Birrarung—along with its law, language, stories, culture, and kin—into the "Yarra River," a "water body separate from the land in which the river is emplaced," stripped of all affective agency.[7] Working in tandem, this colonizing gaze and violence were central to the making of "property" out of Country.

Birrarung as a Confluence of Shared Meaning

After enduring 150 years of mistreatment, and more recently, fragmented river management, Birrarung's true identity—never forgotten by its kin—is beginning to (re)emerge. This (re)emergence has been slow and reflects the confluence of multiple currents. Decades of advocacy for better river protection by those who love the Birrarung/Yarra River, coupled with the initiation of treaty negotiations between First Nations and the State of Victoria, and legal developments overseas have underpinned the creation of the Birrarung Act, which, as Wurundjeri Elder Aunty Alice Kolasa states, "recognises something that we, as the First People, have always known, . . . the Birrarung is one integrated living entity."[8] In doing so, the Birrarung Act marks a shift away from a propertied understanding of the river, a relationship so entrenched in settler legal understandings of land and rivers that to unravel it requires disrupting privileged (propertied) truths.

Disrupting Propertied Truths

> *The image of dead or thoroughly instrumentalised matter feeds human hubris and our earth-destroying fantasies of conquest and consumption.*
>
> —JANE BENNETT, *VIBRANT MATTER: A POLITICAL ECOLOGY OF THINGS*, 2010

A new legal materialism is emerging. The revitalization of "inert" matter to recognize nonhuman agency is challenging long-held, instrumental truths that have shaped and that perpetuate settler legal orders that give primacy to the relationship of property. Helen Verran offers an intervention for acknowledging and actively engaging with contested cultural-legal meanings of place, a practice she calls "doing ontological politics."[9] The work of challenging "property"—a powerful actor—requires the employment of other powerful actors, two of which are stories and legal categories.

Stories

Stories are excellent disruptors. They can do deep ontological work by drawing us into disconcertment (Verran's term), poking holes in present realities and introducing previously unimagined truths. The preamble to the Birrarung Act acknowledges truths that the Wurundjeri have always known: Birrarung is not only living in the sense that it is a river system within a catchment; it is also alive because it has a heart and spirit. Both Birrarung and the Wurundjeri belong to Country; law and Country were created by Bunjil, the great eagle creator spirit; and reciprocity of care guides lawful relationships between Birrarung and its kin. This understanding is reflected in the purpose of the Birrarung Act, which is to protect the river as a relational entity.

The confluence of Dreaming stories and settler law may seem unlikely and uncomfortable. It is yet to be seen the extent to which policy makers and institutions relying on the Birrarung Act can translate this relationship into action that protects Birrarung and its kin. Seeing the river not as property—and not only as a "living waterway" but also as a network of kinship relations that is alive today—demands that I and other settler people reading the Birrarung Act forgo the colonial tendency to relegate Dreaming stories to the "myths and legends" category. It is an invitation to acknowledge Country's agency to be a holder of law, culture, and language. On a deeper level, it requires being vulnerable to allow "one's own [metaphysical] ground . . . [to] become destabilised."[10] While unsettling (and perhaps unraveling) one's reality is deeply disconcerting, it is necessary to allow a new legal materialism to emerge. In turn, this ontological work lays a foundation for productive and respectful engagement between the cultures and legal orders of First Peoples and settler peoples.

Categories

Stories hold and circulate another powerful actor that does deep ontological work: categories. The Birrarung Act recognizes more than *just* a river. By identifying the Yarra River as *Birrarung,* an ancient legal actor and its associations are invoked. It is a coconstitution of human and nonhuman entities: the Wurundjeri First People, their creator spirits Bunjil and Palliyang, Waa, law, language, communities living along the river from its upper reaches to its mouth, a waterway, riverbanks, catchment land, and a story that holds all this together.

By declaring the river as a single, living, integrated natural entity, and by recognizing legal actors like Bunjil, the Birrarung Act takes steps toward recognizing Birrarung as a legal person, an entity with legal rights and obligations that can sue and be sued. Inspired by developments in Aotearoa/New Zealand, Indigenous and settler communities around the world are asking how they might use legal personhood to protect kin and nonhuman nature. While the Birrarung Act consciously stops short of this measure, given the increasing global interest in legal personhood as a means of river protection, it is worth exploring the potential for and limitations of this ubiquitous legal category to disrupt instrumental legal culture and support the recognition of kin.

Legal Personhood

Legal personhood is a powerful and pervasive category that lives and performs across diverse branches of law. The making of legal persons—those who can act in law—is fundamentally an exercise in ontological politics. Peoples enslaved, women, fetuses, sentient nonhumans, rivers, artificial intelligence, future generations, and unitary ecosystems have all struggled to be recognized as persons in settler society's law. Recognition is not and has never been tied to inherent characteristics that qualify entities for "personhood"—it has always been political. For centuries, the declaration of legal persons has been constrained by unquestioned Lockean metaphysics, despite the openness of the category when it was first developed.

Legal personhood can be traced back to the work of Roman scholar Gaius (1904), who classified law into things (*res*), actions (*actiones*), and persons (*persona*). The legal person, or *persona*, represented a right- or duty-bearing entity of some kind. Abstract in nature, *persona* was a legal metaphor drawn from the mask worn by an actor in classical Roman theater. As Ngaire Naffine describes, "artifice or fiction" were employed in the conception of the legal person, demonstrating the malleability of this category and its ability to be applied across and beyond taxonomic classification.[11] Yet the development of the legal person has not always benefited from such broad construction. During the latter part of the seventeenth century, as England emerged out of civil war, John Locke began to articulate the rights of atomized individuals—particularly the rights to life, liberty, and property—against unfair, absolute monarchies. The metaphysics enacted by this incarnation of the legal person has served to perpetuate reductionism in science and positivism in law for more than three centuries.

Personhood in the Anthropocene will, out of necessity, be reflected in myriad different entities. Responding to global ecological crises, human-induced climate change, and a broad failure of environmental laws to remedy these complex and transjurisdictional issues, personhood is one means toward affording legal rights to "natural," nonhuman entities. Novel application of this category may carry with it an explicit mandate, though one that is not guaranteed: to reconfigure human relationships with nonhumans, including person-property and subject-object distinctions.

It remains to be seen whether the expansion of legal personhood to recognize "new" nonhuman natural entities—and kin—will achieve this potential, or whether legal personhood is even needed to decenter the human in dominant cultures in order to redefine what it means to be lawful. After all, personhood—as a category that gives primacy to individual rights holders—does not recognize the collective and relational understandings that are lived through First Peoples' lifeways. Perhaps the relational approach of the Birrarung Act, and with it, the invocation of kin to inform legal relations, will provide an alternative to personhood as a culturally resonant way to protect nonhuman kin.

Palliyang

The sun has long dropped over the lip of the river valley. Dusk has fallen and palliyang rise up from the trees along Birrarung, darkening the sky overhead. I do not know where their nocturnal feeding places lie, how far they travel each night. There is much to learn from palliyang—the bat—and about Palliyang, the creator of Bagarook (women) who came from the waters of Birrarung. Although I have taken only a few small steps in learning how to be lawful with this land, sitting here, next to Birrarung, feels like the right place to allow this question to unfold.

NOTES

1. One of the languages spoken by Yolngu in the Mäpuru homelands is Djambarrpuyngu, which is reflected here.
2. Tyson Yunkaporta, *Sand Talk* (Melbourne: Text Publishing Co., 2019), 169.
3. Ellen Van Neerven, "The Country Is Like a Body," *Right Now* (2015): 1–8, http://rightnow.org.au/essay/the-country-is-like-a-body/.
4. C. F. Black, *The Land Is the Source of the Law* (London: Routledge Cavendish, 2011), 13.
5. *Liyan* is a word commonly used by people across several Nyulnyulan language groups in the North-West Kimberley. For more on its significance in relation to "feeling" Country, see Paddy Roe and Frans Hoogland, "Black and White, a Trail to Understanding," in *Listen to the People, Listen to the Land*, ed. Jim Sinatra and Phin Murphy (Carlton: Melbourne University Press, 1999), 11–30.
6. A songline is an oral heritage "map" that perpetuates the cultural knowledge and Law of cocreation ("Dreamings"). Songs encoded in the land tell people how to be lawful, to live in balance with and be sustained by other nonhuman people, the land, sky, and sea.
7. Cristy Clark, Nia Emmanouil, John Page, and Alessandro Pelizzon, "Can You Hear the Rivers Sing? Legal Personhood, Ontology, and the Nitty-Gritty of Governance," *Ecology Law Quarterly* 45, no. 4 (2018): 825.
8. Auntie Alice Kolasa, Wurundjeri Elder, Address at the Parliament of Victoria (June 22, 2017), https://www.wurundjeri.com.au/wp-content/uploads/pdf/Wurundjeri_parliamentary_speech_download _a.pdf.
9. Helen Verran, "Metaphysics and Learning," *Learning Inquiry* 1, no. 1 (2007): 31–39.
10. Deborah Bird Rose, *Reports from a Wild Country: Ethics for Decolonisation* (Sydney: University of New South Wales Press, 2004), 22.
11. Ngaire Naffine, "Who Are Law's Persons? From Cheshire Cats to Responsible Subjects," *Modern Law Review* 66 (2003): 346–67.

CAN WE GROW THE CONCEPT OF OUR SELVES?

Merlin Sheldrake

From a biological point of view, it is not always clear where the boundaries between individuals should be drawn. A substantial proportion of our genome has been acquired from viruses, and we carry around more bacteria than cells of our own, without which we would not grow and behave as we do. We are ecosystems, composed of—and decomposed by—an ecology of microbes, the significance of which is only now coming to light. Our microbial relationships are about as intimate as any can be, but we are not a special case. Bacteria host smaller bacteria and viruses within them. Even viruses can contain smaller viruses. If the word *cyborg*—short for "cybernetic organism"—describes the fusion of a living organism and a piece of technology, then we, like all other life-forms, are symborgs, or symbiotic organisms, made up of multiple nested selves.

The interconnectedness of our lives may be clearer than ever, but it is still easy to take for granted our concepts of individuality and selfhood. Much of daily life depends on clearly identifiable individuals, after all, as do our philosophical, political, and economic systems. And now we are seen as individual points within epidemiological network models that chart the spread of a virus as it seeps through populations and across borders. Network has become a master concept—in computing, sociology, neuroscience, ecology, and economic systems—not to mention the very structure of the universe itself, now understood to be a vast "cosmic web."[1] Yet just as much as networks reveal interconnectedness, they tell us about how we divide and partition the world. Networks are made up of nodes and links, and it is only once we have identified separate entities that we can string them together.

The life-forms that raise some of the most potent questions about networks, individuality, and selfhood are fungi. Mushrooms are only the fruiting bodies of fungi: for the most part, fungi live their lives as branching, fusing networks of tubular cells known as mycelia. Mycelial networks have no fixed shape. By ceaselessly remodeling themselves, they can explore their surroundings, navigate labyrinths, and solve complex routing problems. For fungi, self can gradually shade off into otherness. A mycelial network can fuse with another network entirely. And a small fragment of mycelium can grow into a new network, which means that a single mycelial individual—if you are brave enough to use that word—is potentially immortal.

Fungi live their lives entangled with other organisms, and mycelium is a living seam by which much of life is stitched into relation. If one teased apart the mycelium found in a teaspoon of soil and laid it end to end, it could stretch anywhere from a hundred meters to ten kilometers. Mycelial networks weave themselves through plant roots and shoots, animal bodies, sulfurous sediments on the ocean floor, grasslands, and forests—one of the largest-known organisms is a mycelial network in Oregon that sprawls over hectares. Bacteria use fungal networks as highways to navigate the bustling wilderness of the soil. Symbiotic mycorrhizal fungi—from the Greek words for fungus (*mykes*) and root (*rhiza*)—grow within and around plant roots and lace out into the soil, where they scavenge for nutrients and water that they exchange for sugars and lipids produced by their plant partners in photosynthesis. This ancient association gave rise to all recognizable life on land, the future of which depends on the continued ability of plants and fungi to form healthy relationships.

Today, more than 90 percent of all plant species depend on mycorrhizal fungi. They are the rule, not the exception: a more fundamental part of planthood than fruit, flowers, leaves, wood, or even roots. Out of this intimate partnership—complete with cooperation, conflict, and competition—plants and mycorrhizal fungi enact a collective flourishing that underpins our past, present, and future. Lynn Margulis described the history of life as a story of the "long-lasting intimacy of strangers."[2] Plants and their mycorrhizal partners are a good example of a long-lasting intimacy, but they are hardly strangers any more. Look inside a root and this is clear.

Roots turn into worlds under a microscope. I have spent weeks immersed in those worlds, sometimes enthralled, sometimes frustrated. Put fresh, fine roots in a dish of water and you will see fungal hyphae stringing off them. Boil roots in dye, squash them onto a glass slide, and you will see an intertwining. Fungal hyphae fork and fuse and erupt within plant cells in a riot of branching filaments. Plant and fungus clasp each other. It is difficult to imagine a more intimate set of poses.

The strangest thing I have seen under a microscope is germinating dust seeds. Dust seeds are the smallest plant seeds in the world. A single seed is just visible to the naked eye, like a small hair or the tip of an eyelash. Orchids make them, as do some other plants. They weigh almost nothing and disperse easily with wind or rain. And they will not germinate until they have met a fungus. I spent a long time trying to catch them in the act. I buried thousands of dust seeds in small bags and dug them up after a few months, hoping that some would have sprouted. Under the microscope, I pushed seeds around a glass dish with a needle, searching for signs of life. After several days, I found what I was looking for. Some seeds had swollen into fleshy clumps tangled up in fungal hyphae, sticky streamers that trailed out into the dish. Inside the developing roots, hyphae raveled into knots and coils. This was not sex: fungal and plant cells hadn't fused and pooled their genetic information. But it was sexy: cells from two different creatures had met, incorporated each other, and were collaborating in the building of a new life. To imagine the future plant as separable from the fungus was absurd.

And yet what I had seen was a simplification. Most plants are promiscuous and can engage with many mycorrhizal partners. Mycorrhizal fungi, too, are promiscuous in their relationships with plants. Separate fungal networks can fuse with each other. The result? Potentially vast, complex, and collaborative systems of shared mycorrhizal networks sometimes known as the "wood wide web," through which water, nutrients, and signals can pass.

Where are the nodes and where are the links? In the networks plotted by the contact-tracing apps being used to track the spread of COVID-19, each link represents just a moment of contact. By contrast, fungi form enduring physical connections between plants. It is the difference between having twenty acquaintances and having twenty acquaintances with whom one shares a circulatory system. From one perspective, plants are the nodes and fungi are the links. From another, fungi are the nodes and plants are the links. In 1845, Alexander von Humboldt described the natural world as a "net-like, entangled fabric."[3] Fungal mycelia make the net and fabric real.

Over the past few decades, as the prevalence of network concepts has grown, so too has a sense of human disconnection—from one another, from the natural world, and from the prospect of a secure future. Network thinking is not the cause of our disconnection, but neither does it seem to be the entire cure. What do we choose to call a node? A vision of society made up of networked individuals is a metaphor that reveals and perhaps reinforces an ingrained individualism. Our individualism shapes the way we form connections with one another and affects the distribution of resources and responsibilities. It is by imagining ourselves as separable—from one another and the ecosystems that sustain us—that we justify both the exploitation and the oppression of other humans and ecological devastation. Pandemics caused by novel pathogens and their unequal impact on different social demographics are predictable consequences.[4]

We may conceive of ourselves as autonomous individuals, but fungi raise questions about our categories in ways that make the world look different. I attended a conference in Panama on tropical microbes, and along with many other researchers spent three days becoming increasingly bewildered by the implications of our studies. Someone got up to talk about a group of plants that produced a certain group of chemicals in their leaves. Until then, the chemicals had been thought of as a defining characteristic of that group of plants. However, it transpired that the chemicals were actually made by fungi that lived in the leaves of the plant. Our idea of the plant had to be redrawn. Another researcher interjected, suggesting that it may not be the fungi living inside the leaf that produced the chemicals, but the bacteria living inside the fungus. Things continued along these lines. After two days, the notion of the individual had deepened and expanded beyond recognition. To talk about individuals made no sense any more. Biology—the study of living organisms—had transformed into ecology: the study of the relationships between living organisms. To compound matters, we understood very little. Graphs of microbial populations projected on a screen had large sections labeled "unknown." I was reminded of the way that modern physicists portray the universe, more than 95 percent of which is described as "dark matter" and "dark energy." Dark matter and energy are dark because we do not know anything about them. This was biological dark matter, dark life.

It made my head spin to think of how many ideas had to be revisited, not least our culturally treasured notions of identity, autonomy, and independence. It is in part this disconcerting feeling that makes advances in the microbial sciences so exciting. Our microbial relationships are about as intimate as any can be. Learning more about these associations changes our experience of our own bodies and the places we inhabit. "We" are ecosystems that span boundaries and transgress categories. Our selves emerge from a complex tangle of relationships that is only now becoming known.

Fungal networks embody the most basic principle of ecology: that of the relationship between organisms. The word *ecology* has its roots in the Greek word *oikos*, meaning "house," "household," or “dwelling place.” Our bodies, like those of all other organisms and the ecosystems we inhabit, are dwelling places. Can we learn to grasp this on an intuitive level? Are we able to loosen the grip of some of our certainties about how we divide the world?

I think so. The more I have studied fungi, the more my expectations have loosened, and the more familiar concepts have started to appear unfamiliar. I now wonder whether our sense of self might be more an assumption than a fact. In the well-known “rubber-hand illusion,” people watching a fake hand being stroked while their own concealed hand is simultaneously stroked become convinced that the fake hand is their own. We are able to radically refigure the boundary of our bodily selves after a mere twenty minutes of experience. What more are we capable of? What would happen if we made individuality and selfhood a question rather than an answer known in advance? Where would this leave “us”? What about “them”? “Me”? “Mine”? Perhaps some of the vexed divisions that underpin modern thought would start to soften. If they did, our ruinous attitudes—toward one another and the more-than-human world—might start to transform.

NOTES

1. Joseph N. Burchett et al., “Revealing the Dark Threads of the Cosmic Web,” *Astrophysical Journal Letters* 891, no. 2 (2020): https://iopscience.iop.org/article/10.3847/2041-8213/ab700c.
2. S. Mazur, “Lynn Margulis: Intimacy of Strangers & Natural Selection,” *Scoop,* March 16, 2009, www.scoop.co.nz/stories/HL0903/S00194/lynn-margulis-intimacy-ofstrangers-natural-selection.htm.
3. Alexander von Humboldt, *Kosmos: Entwurf Einer Physischen Weltbeschreibung* (Stuttgart: J. G. Cotta’schen Buchhandlungen, 1845), 1:33. The sentence containing the phrase “net-like, entangled fabric” (*Eine allgemeine Verkettung, nicht in einfacher linearer Richtung, sondern in netzartig verschlungenem Gewebe, . . . stellt sich allmählich dem forschenden Natursinn dar*) does not occur in the published English translation of *Cosmos* in 1849.
4. Jonathan Watts, “‘Promiscuous Treatment of Nature’ Will Lead to More Pandemics,” *The Guardian,* May 7, 2020, https://www.theguardian.com/environment/2020/may/07/promiscuous-treatment-of-nature-will-lead-to-more-pandemics-scientists?.

BOY I

Heather Swan

He bursts from the cattails
clutching a bullfrog—
the glabrous body
slick with mud,
thick legs outstretched,
but somehow tranquil.
His hands could easily crush
this creature whose soft belly
is the color of milk,
who can breathe
through her skin,
whose only protections
are a transparent eyelid
and quickness.

This is the child who,
in the darkness, unable
to sleep, curls into
the body he came from
and asks, *But who invented war?*
And *Can a bullet go through brick?*
Can a bullet go through steel?

Now, at the water's edge,
filled with a wild holiness,
he navigates the balance,
then lets the frog go.

BOY II

Heather Swan

The green darner unzips
the air above the cattails
and the sparrow threads
the trees with song,
unspooling sounds
that make this landscape
our known one. The smell of
Queen Anne's lace and
juniper on our pant legs,
the softness of a leopard frog's
belly and the weight
of honeybee footsteps
on our hands—
all making a kind of home
for you, for me.
It is to these I turn
when you turn from me
and run—bridle off—
at full gallop into a new field
you will soon know well enough
to call your own.

KINSHIP AND OTHERNESS: THE FINE ART OF SHAPESHIFTING IN MYTH AND FOLKLORE

Sharon Blackie

I've never met a woman who, when the conversation turns that way, doesn't in some way or another identify with old European folktales about part-human, part-animal female beings whose skin is stolen by a man. Think about it: pretend for a moment that you're standing on a white strand in the Western Isles of Scotland, looking out onto a moody sea. It's there, on the night of a full moon, that you might—if you're lucky—catch a glimpse of a selkie. Once a month, she can slip out of her sealskin and dance on the beach with human legs. Stand there a little longer, and you might spot the hungry-eyed fisherman who wants her for his own. So he steals her sealskin, forcing her to stay in human form, depriving her of her sealness. "Stay with me for seven years," he says, "and then if you still want me to, I'll let you go."

Except, of course, he doesn't. And so slowly, trapped in her human form and unable to return to the life-giving sea, she begins to fade. She begins, quite literally, to dry up. On the verge of death, the skin that her husband had hidden away is found by her daughter, and the selkie puts it on and slips back into the sea, back into her natural element.

Selkie tales like this one are told all along the western coasts and islands of Scotland. In these places, the people of the land have always lived alongside the people of the sea. They occupy the same habitat and compete for the same fish; it's not surprising that we find stories in which the two peoples merge into one. Such a commonplace story soon becomes legend, then legend is told as history, and so we find the MacCodrum clan of the isle of North Uist, who became known as the "MacCodrums of the seals." They were descended, it is said, from the union of a fisherman with a selkie; this explained the hereditary horny growth between their fingers that made their hands resemble flippers.

It is no surprise to me that these stories persist so powerfully today. Because I've never met a woman, either, who hasn't at some point in her life felt as if her skin had been stolen. And who among us hasn't felt as if we've been exiled, banished from our natural element, the wild creaturely side of us stripped away to satisfy an excess of civilization? These stories of animal-human shapeshifters are stories of yearning—for a part of ourselves that we feel we have lost—or maybe a part that we feel we might once have had but never properly knew.

In Europe, such shapeshifting stories are ubiquitous. In Croatia, it's the story of a wolf woman whose pelt is stolen by a hunter. In Ireland and Sweden, it's a swan maiden whose feathers are stolen by a covetous man. But the union between human man and shapeshifting animal-woman isn't always forced: countless stories have been recorded in which an animal shapeshifts into human form so that she can voluntarily enter into a relationship with a man. These are known as "animal bride" stories. A man wins her love, they marry, and for a while, they live happily. They have children. One day, the husband makes a mistake or violates a prohibition she has laid upon him; the wife reverts to her animal form and leaves, sometimes (but interestingly, not always) taking the children with her.

In a few of these stories, the animal brides are rather more ambiguous creatures. Just like the not-to-be-messed-with fox wives of Japan and Korea, the forest-dwelling, foxlike huldra of Swedish lore tends to be a beautiful, dangerous being who slowly drains the life energy from her human lover. Sometimes, these stories remind us that what is wild is not benign. And what is wild is not to be exploited; rather, if we're not careful, it will exploit us. This aspect of our native folklore reassures me because it seems *real*. It reflects a world in which shadow is always found at the heart of the light, in which actions have consequences and, sometimes, a wild animal will tear your heart out just because she or he can. "Don't take anything for granted," these stories whisper; "not everything is about you." You really aren't master of the universe; you're not even mistress of the beasts. So above all, pay your respects.

In many of these old shapeshifting tales, the husband symbolizes the human world while his animal wife represents the natural world. The persistence of these stories into modern times, then, can be seen—in addition to representing the dangers of disrespect—as a reflection of our longing to reestablish a lost relationship with the natural world. And changes in the tales, as they pass through the centuries and transform themselves (as all folktales inevitably do), reflect that changing relationship between humans and the natural world.

The oldest recorded tales of animal brides, then, include ancestral myths, such as those Scottish island stories of families who claim descent from the marriage of humans and shapeshifting seals, or of Siberian shamans who trace their power back to marriages between men and swans. In some cultures, marriage between humans and animals might break certain taboos, but in the oldest tales these relationships were rarely portrayed as evil or immoral. And even when the marriages were doomed to failure, as so many of them were, more often than not a gift was left behind, in the form of children, wealth, or magical abilities.

These early traditions more clearly reflect a pre-"Enlightenment" worldview: one in which humans were viewed as an intrinsic part of the natural world, as related to the animals—rather than being separate from nature and possessing dominion over all the other creatures within it, as so many of us imagine ourselves to be today. But later stories of animal brides are much more likely to end badly—arising, no doubt, from a worldview that perceives humanity and its "civilizations" as both distinct from and superior to nature and the wilderness. Humans and their animal lovers come from decidedly separate worlds, these stories suggest—and any attempt to unite the two must ultimately, and rightly, be doomed to failure.

So in Europe, by the medieval period, we find that creatures who could shift between human and animal forms were portrayed in distinctly demonic terms. One of the best-known such tales is the French story of the fairy Mélusine, which was first written down in 1211. A count called Raimondin meets the beautiful, otherworldly Mélusine beside a fountain in a forest and falls in love with her. She agrees to marry him on one condition: he is never to see her on a Saturday, when she always spends the day bathing. He agrees; they wed, and she bears the count nine sons—each one of whom is deformed in some fashion. After many years have passed, suddenly suspicious of her, Raimondin breaks the taboo and spies on her while she's in her bathhouse. And so he discovers that, every Saturday, she becomes a serpent from the waist down. He is appalled. When she finds out that he has broken his promise and disrespected the one mystery that she wanted to keep for herself alone, Mélusine vanishes.

It is interesting that stories of animal brides portray the brides' animal and human natures as inextricably intertwined; they are both animal and human at the same time, and their animal nature (represented, for example, by the selkie's skin) cannot be taken away from them without running the risk of killing them. This sits in stark contrast to stories in which it is a bridegroom, rather than a bride, who takes on animal form, which is usually portrayed as a punishment or a curse. In the well-known folktales "The Black Bull of Norroway" and "East o' the Moon, West o' the Sun," the male protagonist—a prince, of course—has been enchanted. In the former, he has been turned into a bull; in the latter, into a bear. Often in such stories, the transformation has been inflicted on the prince because he has refused to marry a wicked woman, who then becomes his adversary. It's a woman—who ends up as his bride for her pains—who must break the curse for him and so restore him to his natural human form. If we compare these two different story traditions, then, we can conclude that it was perceived to be more "natural," it seems, for women to become animal than it was for men. The women remained animals; the men returned to their human forms.

Right up to the present day, modern retellings of these old tales, not surprisingly, usually reflect a rather different perspective on human-animal relationships. And it should be noted that there are many such retellings: since the middle of the twentieth century, it has been quite common for writers of fiction to subvert traditional fairy tales, retelling them to reflect contemporary concerns and situations—such as the groundbreaking feminist retellings of writers like Angela Carter.

In *Foxfire, Wolfskin*—my recent collection of fairy-tale reimaginings about shapeshifting women—I use animal transformation motifs to explore the qualities that might lie at the heart of women's creative power today. The stories are not just about women's remarkable ability to transform themselves in times of crisis—but they are about the ways in which our innate association with the wild, and our kinship with nature, is reflected in a time of environmental crisis. These are, in a sense, stories of resistance—made for a time when we are living through the overturning of many aspects of the traditional social order. Can such stories, I wanted to ask in that collection, help us renegotiate and reimagine our fractured relationship with the natural world today? They certainly reflect my longing for a wildness that is very far removed now from our daily lives—including a yearning for now-extinct creatures (in the lands of my ancestry, the wolves and bears are long gone, although they still exist in other parts of the world), which live on only in our imagination. As well as representing the lost or repressed personal Wild Woman (or Wild Man) archetype that our culture finds unacceptably nonconformist, these shapeshifting beings represent a rich and necessary wildness that our civilization has all but wiped out.

I believe—and perhaps it's not surprising, since I originally trained as a psychologist—that the major reason for the enduring popularity of these old shapeshifting stories is quite simple: at the heart of all fairy tales is *transformation*. The tales help us to reimagine ourselves. They help us not only to unravel who we are but also to work out who we want to become. The many-layered images and the fragments of hope and yearning that are embedded in these old tales clearly demonstrate the essential power of story: they breathe life into our thoughts and dreams; they show us how we might possibly play our own unique part in the ongoing becoming of the world. Fairy tales, from a very early age, teach us the fundamentals of what it is to be truly alive. Their lessons are deep and rich. Anywhere there may be a door to another world: learn to look for it. Always leave a trail of bread crumbs to find your way out of the dark wood. Don't maim yourself trying to fit into the glass slipper that was made for someone else. Even though you're hungry, give half of your porridge to the mice who might help you sort the wheat from the chaff. And never—ever—take your skin off and leave it unattended.

Fairy tales, then, declare that there is more to life than we can ever possibly know, and they reveal to us glimpses of the mysteries that lie behind the particular aspects of the physical world we choose to focus on—mysteries that usually are hidden from us. Stories about shapeshifters, about the fluid relationship between human and animal, between "civilized" and wild, help us come to see that there are other ways of imagining the world and our place in it, and they show us how to live more intensely, and more richly, in that multilayered and suddenly animate world.

Such a shift in perception happened to me when I moved to an old stone cottage in Ireland where we lived by a river. At the back of our house was a heronry. In the Irish language, the word for the gray heron is *corr*, which also happens to be the word for *crane*. This is because, just around the time that the Eurasian crane became extinct in Ireland, the similar-looking gray heron arrived to fill its ecological niche. Heron and crane, then, are interchangeable in Irish mythology, and in the old stories, they are powerful and liminal birds. She haunts the thresholds where water, land, and air intermingle; she guards the treasures of the Otherworld and is a guide to those who wish to travel there. Perhaps because she stands upright, tall and thin, she is associated with shapeshifting in the feminine form (mostly shifting into old women)—and indeed, most likely for this reason, eating a crane's or a heron's flesh was once forbidden.

So I knew what a heron was in my mythology. But every morning, startling a heron up from the riverbank as I walked by, listening to her haglike shriek as she took off, I wondered as she circled around and looked back down at me, *What am I, though, in the mythology of a heron? What does heron see when she looks at me?* And then—*what is it that I would like her to see?* So it is, I believe, that these old stories about animal shapeshifters teach us to think about both perspectives, to interrogate the world with both human and animal eyes. I look up at heron, looking down at me, and somehow the boundaries between us blur. It's easy, then, to imagine a kinship; it's easy to slide into relationship, to talk to a heron sitting on the riverbank, to sing her a song or offer her a little poem, to court her friendship just as I might court the friendship of a fellow human.

Although they teach us to see the world, and the other-than-human creatures who inhabit it alongside us, as alive and sentient, stories of animal brides in particular—especially those in which marriages between animal and human falter or fail—also teach us about how difficult it can be to hold the balance between kinship and otherness—that is, how to navigate the balance between culture (the human husband) and nature (the animal bride). Although these stories might affirm our deeply entangled relationships with animals, they also force us to acknowledge their strangeness, and—if we allow ourselves to fully inhabit the imaginal world of the story—to see this ever-shifting balance as a mystery, which is always to be inhabited, rather than as a problem to be solved.

Ultimately, even the most cursory readings of the old myths and stories of Britain and Ireland make clear that our ancestors believed that the borderlines between humans and other animals were much less clearly defined than we imagine them to be today, and also more easily bridged. These tales had their origins in times not only when there were closer connections between humans and the natural world but also when there was no differentiation between secular and sacred—the sacred existed everywhere, in everything. You didn't need a special building to find it or a special person to act as your intermediary; it was right there, all around you—in that tree, or that stone, or that fish. In those times, then, animals were very much more than simply creatures that we might hunt and eat: they were teachers, and allies, and sometimes they were even gods. Some of those oldest stories involve powerful elder animals who embody a kind of wisdom that is different from (but entirely congruent with) the ways of knowing that humans have. Animals, of course, can do things that we can't, and they can sense things that we don't know about—and in these old tales, wild animals in particular were repositories for the kind of knowledge that we long ago renounced.

The French philosopher René Guénon once suggested that we live in "degenerate times," at the end of a long age during which important spiritual truths have been forgotten, the ancient centers of wisdom have been destroyed, and the guardians of that wisdom have been dispersed. At such times, he said, a safe repository for spiritual truth can be found in folklore. He suggested that knowledge that is in danger of being lost passes into the symbolic code of a folktale and so passes on to the people. They will perhaps be concerned only with the stories' surface meanings—but they will at least preserve them and pass them down to their children. Then, in better times, people might once again appear who will try to penetrate the symbolic disguise to discover the stories' wider meaning.

Whether or not they contain the encapsulated wisdom of ages past, what is certain is that myths, fairy tales, and folklore offer us a world imbued with participation mystique—a world in which humans are fully enmeshed. In this world, animals always have something to teach us or give us. It is profoundly important that we revisit these old stories, because the stories that civilizations tell about themselves and their relationship with the world around them are the stories that determine the overall shape of individual and communal lives. They determine our worldviews: the assumptions we hold about why things are the way they are, about how things work, about what is valuable, and about what is possible. And in turn, these assumptions underlie the choices we make, large and small, every day of our lives.

Approaching fairy tales and folktales in non-human-centric ways places us more firmly into the wider life of the world: our personal story is entangled with a greater story of which we're a part. That sense of awe, of connection, of belonging to a mysterious, fully inhabited, animate world, which has many depths and layers to explore, is precisely what humankind needs today.

A LANGUAGE OF LISTENING

Julian Hoffman

Still the memory glimmers. Over twenty years later, and half a world away from the forest and plains where the elephant dwelled, it feels as though the experience has been somehow isolated from others, so that it shines as bright as a lone fire at night. A fire that's been shored up and sandbagged against any floods of forgetting to come.

It was a blue-skied and sweltering afternoon when my wife-to-be and I saw taped to a bamboo shack the handwritten sign offering elephant rides. Having only recently acted on our mutual attraction to each other, we were backpacking through the temple towns, backwaters, and jungles of southern India, and so the idea of sitting together astride an elephant in the simmering light of Kerala, seeing this new and beckoning land from the unique perspective of the elephant's back at the same time that we urgently learned each other, seemed impossibly wondrous to us.

We were led to a forest pool where the elephant was being bathed by its keeper, shoals of waterfalling light sliding from its stippled gray skin. I had never before been in the presence of such an overwhelming sentient force, and the dense forest stillness seemed to coat the moment like snow. Placidly vacuuming up water with its trunk to sluice clean the great fans of its ears, the elephant jetted a net of sparkling spray high into the air above us. Finally ready, the keeper called a command. The elephant promptly raised the pillar of a back leg for us to hitch a lift on, ratcheting us upward onto wide leathery slopes to summit more easily the stone-colored ridge of the elephant's back. Once mounted, high and exhilarated above the ground, we moved through the forest with ease, roving slowly through the sieved light of trees. Elated, we tracked the edge of an earthen village where children waved their hands as we passed, and then we pushed on into open country.

Insects rose in great swells from the pale grasses unrolling into the distance. We scattered birds from low-hanging trees, our heads raking through leaves as we bent forward to better shape ourselves to the great being beneath us. Energy was palpable through skin, something elemental and old, something as transfixing and fizzing as the spark from crossed wires. For the next hour we were carried as if kings, and all the world seemed to be ours.

We say that elephants never forget, but what we mean by this is that elephants remember. They remember ancestral paths across plains that look trackless to us. They remember bevels in the earth that briefly brim. They remember seasons of hardship and seasons of prosperity. They remember each and every member of their herd, or even elephants they have not seen for many years. They remember where rain seeps away, where rain makes rivers. They remember human languages, ages, and genders, as revealed in a study in which elephants reacted defensively to the sound of adult, male Maasai speakers, who, as a people, have been known to kill elephants, while ignoring the voices of young boys and women in the tribe, as well as those of the Kamba, a people who do not hunt elephants. And elephants remember the dead.

Some scientists and anthropologists believe that elephants mourn their kin by ritually responding to found bones, raising their legs over them in an unusual display and tenderly touching them with their trunks, something that never occurs with the carcasses of other species. It is thought that their reaction to the bodies of brethren animals reveals a comprehension of mortality, a self-awareness and conscious acknowledgment of life's limited parameters. It is also thought that this grieving for the remains of others is emotional, empathic, and connective in quality.

But what of piano keys, jewelry, furniture inlay, and billiard balls? What of chess pieces and decorative figurines, carvings to enhance one's status, trophies displayed on a wall, and elaborately scaled replicas of castles and sailing ships? What of pale powder mounded up in jars in a shop of traditional medicines in a Chinese city? Would elephants recognize the transfigured tusks of their kind and still touch them with their trunks, or has ivory been so successfully shapeshifted by us that it would be alien even to those it belongs to?

Between ten thousand and fifteen thousand elephants each year are purposefully killed by poachers in Africa alone, which means there are now more bones and bodies of elephants to grieve over than ever.

Far too many to forget; too many to ever remember.

Nearly two decades later, my wife and I returned to India. One evening, as we sat around a bamboo campfire that crackled outside a tribal home in the jungles of Arunachal Pradesh, we met Rinku Gohain. Laidback, engaging, and self-deprecatingly funny, Rinku was a veterinarian for the Wildlife Trust of India. Based at the nearby Pakke Tiger Reserve, one of Rinku's duties included the rehabilitation of Asiatic black bears that had been intentionally harmed or accidentally injured. The work of reintroducing them to the wild, often after long periods of convalescence, sometimes required him to spend the monsoon season on a swaying tree platform in a rain forest literally dripping with leeches to make sure the bears readjusted. And there he stayed amid storms, alone except for his rewilding charges on the forest floor below him, until the monsoon rivers receded sufficiently for him to leave the jungle, sure in the knowledge that the bears he'd cared for were at home once more in the wild.

Although I'd only just met him, I immediately admired Rinku. I held his tireless efforts on behalf of wild animals in deep respect, and his thoughts on elephants—the animal, it soon became clear, was not only his specialty but also the driving force in his life—moved me considerably. This particular path of his began when he helped out in a sanctuary for working elephants. The veterinarian in charge was using massage therapy to treat domesticated pachyderms that had come from logging camps, temples, tourist centers, agricultural settlements, and forestry departments. Until then, Rinku had stuck to a more conventional practice as a veterinarian in his career.

"I had been understanding elephants in the wrong way," he said matter-of-factly beside the fire. "In the regular way, the way that people had been telling me to for a long time."

Being at the sanctuary, though, opened Rinku to something profoundly unexpected in the lives of elephants.

"I was only there as an ordinary vet at first, checking blood pressure and monitoring reactions, but I ended up becoming an animal behaviorist when I started noticing everything I saw as changes," he said. Rinku coaxed flames from the fire by pushing a length of bamboo deep into the glowing coals.

"When a pregnant elephant, or a stressed elephant, is there, and someone is giving it a massage, you see all the changes that take place until the animal is totally relaxed. The elephant will fart; it will start snoring. These are all good signs. So I began thinking about it, going deeper and deeper into the issues, and reading about it, too. And then I saw a whole different angle emerge: that in India we have understood everything the wrong way about elephants. Obviously, in ancient times, when there were no tanks or vehicles, people were bound to use elephants as their strongest vehicle, or the armor of the army. But times have changed. People don't need to see elephants in the same way as they did in the past, but unfortunately people still understand elephants in the same way. They still think they can *use* elephants because of their strength, but in doing so we are neglecting the whole thing: what elephants are and what they should be," he told us.

Rinku paused and stirred the fire with a stick: "Training breaks the mind. It means we are making a person totally give up everything in life and become like a slave to a master. So it's the breaking of the mind, like we are mentally breaking them, making them ill. Elephants still have the capability to feel and regrow their whole wildness, but we are still not allowing this. In India we pray to elephants, to Ganesh. We consider the elephant a god, and then we chain the god."

Listen, I say to myself. Spool back time to that Keralan plain again. Now cup your ears. *Yes, I remember.*

Those insects, their sound as thick as treacle in the grasses: sawing, clicking, wheezing, churring. The lone drone of a barbet in a tree. The children's high, excited voices when they waved, although I don't recall their words. The clapping of wings from those scattered birds. Our laughter and love. Kingfisher calls like the sharp edge of stones. The hum of heat. The commands of the keeper.

And of the elephant?

I remember only silence.

We've long known elephants to be highly vocal as well as social creatures, but it wasn't until the mid-1980s, through the work of the scientist Katy Payne, that we learned just how *little* we knew about their communication system. At a zoo in Portland, Oregon, while watching a pair of Asian elephants one day, Payne suddenly felt vibrations in the air, although there was no corresponding sound that she could discern. Intrigued, she went on to eventually discover that elephants regularly employ an infrasonic register when communicating with one another. As these low-frequency noises are typically beneath 20 Hertz, they exist below the threshold of human hearing—and that led to these sonic resonances being described by commentators as the secret language of elephants.

The Elephant Listening Project, founded by Payne in 1999 and operating out of the Cornell University Laboratory of Ornithology, documents the complex richness of elephant conversations, particularly those of forest elephants in Central Africa. This work, which utilizes an array of sophisticated recording equipment to capture the bioacoustical environment in all its subtle entirety, has revealed that these low-frequency sounds can roll like invisible waves for several kilometers, able to be picked up by the feet of other elephants as well as their ears, making conversation possible within dense jungle where visibility is strictly limited, or across vast open plains where it's easy to get lost amid all that space.

The project has also disclosed that in the Central African Republic as much as 60 percent of an elephant's communications are infrasonic, or inaudible to the human ear. Which means we aren't even capable of registering, without significant technological assistance, more than half the sounds of Earth's largest and most obvious land animal, let alone understand them. Which begs the question, What are we missing when it comes to the rest of life on the planet?

"We're in kindergarten," said one of the scientists about the listening project's discoveries. "We're just learning the very first few words."

By painstakingly marrying elephant behavior observed over the course of years to the digitally recorded sounds, the researchers are compiling what they describe as a dictionary, using these correspondences to tease out the inner meanings of the language.

The flames from the bamboo fire had lowered to a sunset glow by the time Rinku began discussing a domesticated elephant called Bijaya. Local villagers had described her as erratic after she killed a drunken man who got too close to her.

"This elephant," said Rinku, "had three tragic moments in her life. Bijaya had a calf. They took the calf away from her, right before her eyes. So she had that stress. Second, her sister died right in front of her. Third, her mother died right in front of her, too. So, this elephant had three traumatic moments in her life. And so she's completely lost it now. People say that she's a killer elephant, that she's very unpredictable, that she's this and that, but they haven't ever given a thought to *her* side of the story. They have taken away her child, right in front of her. Her sister died, which meant she lost a very good friend, as they used to stick together all the time. Then her mother was the only companion she was left with. So she was bound to lose all hope in life, everything in life that she had."

And if we were to compile a human-to-elephant dictionary to complement the other, what first few words would it be best for elephants to learn?

Beauty, poaching, pain, sentience, ivory, reverence, chains?

On that second visit to India, we'd been considering another elephant ride. It was a desire born of that wondrous, remembered closeness—what had felt like the compelling proximity of the wild. Deep down, though, I knew it wasn't the wild we'd experienced. That part was a mirage, a shallow gloss on the surface of the ecstatic encounter. Any remaining ancestral connection to that original, untempered state had been erased, or it was buried unreachably deep in the core of the elephant. The elephant's life had been transformed by the savage rigors of training—"it's the breaking of the mind, like we are mentally breaking them, making them ill," Rinku had said—turning a wild animal into a large, controllable commodity.

Rinku was unaware that we'd talked about riding an elephant again. His words weren't meant to be tutelary in any way; he was simply passing on what he'd learned in life, the kind of understanding that is often born of a long and transformative personal journey. Instead, he opened a door for us with his moral clarity that evening, giving us space to step through on our own.

We knew our decision was a small thing when compared to the many threats that wild elephants face, but that fire-lit conversation, when smoke drifted upward into the jungle night and a dew of stars glazed the darkness, enlarged one of the fundamental shifts in perception that is required of us not only to acknowledge the wild world as kin but also to ensure the viability of our shared futures: making respectful the relationships that hold us together in place.

One path is through listening. Listening not only to those, like Rinku, who have made it their life's work to care in some way for the wild world, gaining hard-earned and precious insight in the process, but also listening to our wild kin themselves, from salamanders and fireflies to big bluestem and whales. Our obsession with human exceptionalism has meant, for most of the planet's major cultures, that we prefer speaking to listening when it comes to the more-than-human world, defining our connections not through affiliation but through hierarchy, dominion, and control. Even the description of elephants' infrasonic language as secret betrays the believed centrality of humans, as though the language, like those of countless other species conversing in ways we are unattuned to, were somehow being kept from us, despite being openly comprehensible to the animals it belongs to and is needed by.

The real secret is in learning a new language of listening. A language founded on association, empathy, and affinity; a language shaped to the needs of multiple species so that it functions not to exclude from the conversation but to enlarge our idea and understanding of relationship itself, taking into account the composite world we are not only a part of but entirely reliant on; a language in which we're able to hear the *other* side to Bijaya's story, internalizing the pain, suffering, and vulnerability of fellow beings with the concern we feel for our own—a language that asks us, above all else, to be silent at times.

> *Language can only deal meaningfully with a special, restricted segment of reality. The rest, and it is presumably the much larger part, is silence.*
>
> —GEORGE STEINER, *LANGUAGE AND SILENCE*

Silence:
a muting of voices,
or a space given to hearing them.

Before leaving India, we explored a national park with a guide and jeep. As day turned to dusk over the rolling flatlands alongside the Brahmaputra River, park staff approached on the back of an elephant. The animal's eyes had been ringed with black greasepaint to distinguish her from her wild relatives, which lent her a haunted look in the half light. Slowing of her own accord as she neared our open-topped vehicle, the elephant unrolled her trunk to feel my wife's face. The movements were tender and restrained, as if she were searching for something breakable in a darkened place. She read the encounter through caress: the intimate grammar of skin, the subtle vocabulary of touch. As she felt her way into understanding, gently tracing a language across the face of another, I heard my wife crying lightly beside me. And then her whispering "I'm sorry" to the elephant in reply to the things she had shared.

I peered up into her ringed eyes, their watery shine still visible as dusk dimmed the world, wondering what of the wild might still be inside. *We say elephants never forget, but what we mean by this is that elephants remember.*

The mahout called a command and our ways parted like a split river. Moving off into the gathering dark, the elephant left us in silence on the plain.

A silence not empty but swelling.

FINDING KIN IN MADAGASCAR

Eleanor Sterling

I had never played video games prior to the fall and winter of 1992, but I then suddenly found myself—crouched over the screen of my new desktop computer—playing games mindlessly every minute I was not obligated to serve as a teaching assistant.

It was so unlike me to avoid writing my doctoral thesis this way. I had always been studious, even taking my history textbook to a rare family dinner out at Larry Blake's Rathskeller restaurant in Davis, California, for fear of not being prepared for an upcoming test in high school. In 1991, I returned from fieldwork in Madagascar and dutifully got to work on data transcription and analyses, printing out page after page on a daisy-wheel printer and poring over the results. Creating table after table and graph after graph, working to achieve the requested precision and detail, I had outlines for all my chapters and was ready to finish. But I could not write.

Thinking that my problem was having to compose text on the communal computer in the Anthropology Department offices on Hillhouse Avenue, I took out a student loan to purchase a desktop computer, and I excitedly unpacked the white box with large, black, bovine spots. By working at home, I reasoned, surely I could finally concentrate and finish. Instead, I was obsessively playing Minesweeper. For hours at a time. Over months. To the point that my flatmate, Greg, was considering an intervention.

For my doctoral thesis work, I had lived for two years in a tent on a small uninhabited island off the east coast of Madagascar studying one of the world's rarest primates, the aye-aye. The aye-aye is a kind of lemur found only in Madagascar and one of the most unusual animals on the earth. A constellation of unique characteristics—including batlike ears; a foxlike tail; a chickenlike third eyelid; bird-feet-like, long, filiform middle fingers with no flesh; rodentlike continuously growing front teeth; and mammary glands on their bellies—set this creature apart from all other primates. The aye-aye's unusual morphology had attracted Western scientists since the late 1700s, inspiring controversy about whether aye-ayes are most closely related to squirrels, tarsiers, or even kangaroos. The anatomist Richard Owen (famous for having coined the term *dinosaur*) finally undertook the definitive anatomical study, placing the aye-aye firmly within the primate taxonomic order. Having quelled the century-long debate over the aye-aye's taxonomy, Owen was so impressed by the species that he wrote a piece, three years after the publication of Darwin's *On the Origin of Species,* detailing how the aye-aye's perfect synergy of unusual morphology and presumed behavior proved that natural selection could not have taken place and only God could have created such a creature.

At the time of my research, 125 years later, there had been no long-term studies of the behavior and ecology of this enigmatic, nocturnal animal. I had naïvely decided that it was time to learn more about this species and that such research could inform conservation efforts.

The island where I undertook the study, Nosy Mangabe in the Bay of Antongil, is a gorgeous, steep-sloped, forested nature reserve, complete with a sparkling waterfall right next to camp. Idyllic for short visits, it posed challenges as a multiple-year research site. Average annual rainfall is more than 4,000 millimeters, or thirteen feet, and I soon learned that everything molds in this climate (passports, money, clothing, paper). There was no electricity, so the Malagasy rangers and I followed animals all night with flashlights. I had a portable solar panel for rechargeable batteries, but frequent clouds and rain impeded charging and meant we often ran out of power in the middle of the night on the far side of the island and had to wait for daylight to make our way home. Climbing down slippery, steep slopes at night meant the last-ditch grabbing of hanging vines for support, vines that too frequently harbored stinging scorpions, leaving hands swollen and painful for days (the smaller scorpions packed the most pain). Working in the dark and wet meant that the ubiquitous leeches would find their way into our socks, suspend themselves from our binoculars, and crawl into our mouths and evade our fingers that were too soggy to pull them off. Finding and putting radio collars on solitary, secretive animals that nest high up in trees took months; habituating the animals so they did not run away as we were following them took another few months. It was exhausting to work all night and not be able to sleep in the scorching heat during the day.

Yet I recall, as if it were yesterday, the elation I felt the night thirty years ago when I first realized we could start taking data. It was pouring rain and the rangers, Victor and Barthélémy, were asking me over and over, "Where is the animal we are following?" The fancy (and heavy) telemetry equipment I was carrying let me know we were relatively close, but the acoustics of the place echoed such that it seemed like the sound in my headphones was coming from a submarine far away. I closed my eyes to concentrate and turned a circle with the double H-shaped antenna and then sighed in frustration as it caught, as usual, on a hanging vine. I opened my eyes thinking to untangle the vine, only to find that the antenna was instead impeded by a tree trunk just where the animal we were following was feeding. I was less than two arm's lengths away, and the aye-aye, Namana, was busy concentrating on the food and ignoring us. Clearly, we were not impacting his behavior and could begin taking data on his activities.

Therein followed eighteen months of intense experiences—getting to know the handful of animals with radio collars, their nightly patterns, their interactions with each other and other species, and also learning the 520-hectare island so well that we could make our way home in the dark when the flashlights failed just by sensing the slope of the mountainside. We would gear up in the late afternoon, trudge to the base of the tree harboring the nest of the animal we were to follow that night, dig our heels in for perch on the precipitous slopes, and lean back to watch as she or he slowly woke up after dark, hanging upside down from a branch, fluffing her tail, scratching his fur, cleaning her face, before finally setting out for the night.

The animals walked and walked and walked through the night, lithely jumping from branch to branch in the canopy while we scurried up and down slopes below trying to keep up while taking notes. We devised a system of intersecting flashlight beams that allowed the three of us following an animal to keep him or her in the light. One person would keep a light on the aye-aye from below, and someone else would anticipate where she or he might go, race forward, and then train the flashlight behind to catch the animal as she or he moved into the light, all the while avoiding catching the animal's face in direct lighting that might bother sensitive eyes. We even invented a series of hoot calls that communicated where we were going and what the aye-aye was doing.

Through hard work and perseverance, we learned so much. We were the first researchers to see aye-ayes mating in the wild. We learned that aye-ayes eat far more diverse foods than originally thought. We learned that while one animal will work industriously to build or refurbish a nest and sleep in it for days at a time, other animals will also use that same nest at other times. The latter discovery was at once intriguing and disappointing because researchers had hoped for a predictable pattern to aye-aye nest building to be able to use nests as a proxy for determining population sizes of these difficult-to-locate animals. Our finding of communal and unpredictable nest use sank that hope.

There is something about being out at night that heightens your senses. The rangers and I would frequently hear unusual sounds and whisper back and forth in the dark, hypothesizing about what the animal was doing to create that noise. Frequently, aye-ayes too high up in trees for us to see would chuck food remains down. Bits and pieces of fruit and twigs would rain on our heads, and we would scramble to find them on the leaf litter and piece together what the animals were doing.

We named the aye-ayes we put collars on. One, mentioned previously, we called Namana, which means "friend" in Malagasy. Because he was the only aye-aye with a radio collar, we followed him night after night for months, causing me to wonder whether I was going to have to write my thesis on *an* aye-aye of Nosy Mangabe. All that time together allowed all of us to develop an intense understanding of him and his quirks. One time we heard a strange scratching sound. We crept closer, only to find Namana clinging to the side of a tree, placing his nose adjacent to the trunk, curling large ears forward, then flicking fingers quickly in succession and looking up in the air. We looked at one another, puzzled, and inched closer, crouching low to avoid disturbing him. We spotted a trail of ants streaming up the trunk. Namana was flicking them into his mouth from a distance with his finger, similar to people tossing popcorn in the air and catching it in their mouths! We never saw other aye-ayes doing this.

We kept identifying males, and as we learned about them, we recognized that each had his own personality and habits. But we wondered where the females were. When we finally found and tagged a female, we called her Nirina—"loved one" in Betsimisaraka, the Malagasy dialect spoken where I worked—because we had spent so many months wishing for her.

We often pondered whether the animals just took us for random forest beings. Then one day, Louise, a friend from college who had come all the way from the United States to visit and learn, came out with us on our nightly study. We sat at the base of a tree with a nest and waited for Fotsyave to come out. He emerged slowly, as usual, and went through a daily set of ablutions ("faire la toilette," as Victor called it), hung face down from the tree limb, and then suddenly and unexpectedly started climbing down a vine toward us, heading straight for Louise. She said nervously, "What's he doing?" Victor, Barthélémy, and I shrugged, effectively communicating, "We don't know. We have never seen this behavior before." Fotsyave completely ignored Victor, Barthélémy, and myself, and stopped in front of Louise, cocking his head from side to side and sniffing loudly. After a few minutes, he headed back up into the canopy and on his way. We thereafter saw this behavior only in the rare instances when new people came with us—despite the dark and the distance, the aye-ayes were clearly able to distinguish the researchers to whom they had become accustomed from those who were novelties to them. This made me feel, in some way, a part of this family of animals who know their small island home and the diverse beings on it.

I did eventually finish that dissertation. After reflecting with Louise about why I was incessantly procrastinating with Minesweeper, I was finally able to articulate the problem. The disembodied, authoritative, Western scientific lexicon was one way to describe the aye-ayes, but it did not and could not convey the richness of the experiences I had while learning about them. I felt somehow that if I finished my thesis as proscribed, I would be doing a disservice to the magnificent, complex animals I had studied and the place where they lived. The scientific results were part of the story, but not the full story. No table or graph would sum them up. No statistical analysis could approximate their lives.

In the spring of 1993, having receiving two job offers, both contingent on my finishing my PhD, I did what I needed to and wrote everything up and moved on. Until now, I have never written what my most vivid memories and experiences were, never mentioned what was most formative, never said *why* aye-ayes matter to me.

I have been a resident of New York City for two decades. But I still dream about those times in Madagascar and the experiences we all shared together. I wake up missing the thrill of waiting in the dark for an aye-aye to emerge from a nest, watching one of them hang upside down while grooming his or her tail, and then look toward us and swing up onto a branch to start a nightly adventure. The ache is almost visceral, knowing I will never have those experiences again. My research in Madagascar occurred decades ago, yet I continue to carry those relationships with me wherever I go.

SOMEWHERE AMONG OTHERS

Manon Voice

In the wooded den of a local park
Named by a man who has not lived

Half the lifetimes of its giant dusky-limbed ancestors,
Barren tree colonnades circumscribe the shorn-footed paths,

Flinging wide their aisles in trust.
This place has been named passive

For its low development and I cannot help
But wonder over how it yields for us.

Apart from the worlds we make in our minds
There is right here another more primordial one

Growing alongside the vast tracts of highways
Burrowing through the passages of deep time.

Here, where these sylvan sisters rejoice in their Spring nudity
Shy but never ashamed.

No matter the dermis of flayed tree bark
Swinging in the shaded wind, disclosing its hollowed entrails,

No matter the newcomers with calorie tracking devices,
Oldcomers with their slow and lowly gate, no matter who
does or does not take notice.

Here, are the braided vines sprawling atop the emerald campus
Of early shrubs bulbing their diminutive heads.

Here, is the scalped crown of timber bearing an effigy
Of a wounded lord of nature omitted of hands

Watching over this earthly haven
Of bitter soil and darkened wood

And other countless slain, vouchsafing their cambium
Pulverized to rubiginous dust.

Here, contiguous trunk plinths rounded as requiems
And lattices of wooden crucifixes,

And a bonfire made of thick limbs above the creek,
The timorous flow of water beneath.

Here, the towers of oaks and lindens bearing
Their fresh foliage as wreaths,

The interweaving of hands,
The mothers who push their wisdom into the plexus of
their young

In the community of life, death and rebirth,
And giving and giving away.

And in all the world making for itself a name
Praises for the anonymity of this ecology,

The life that is satisfied to move in the night
And make a celebration of endarkenment.

And of all those in the world making for themselves names,
Have I ever dared to be unknown; to be one simply satisfied
with belonging among the many of others?

A LITTLE MORE THAN KIN

Richard Powers

Reflecting on a possible genetic basis for altruism, the evolutionary biologist W. D. Hamilton once wrote that "in the world of our model organisms . . . everyone would sacrifice [his own life] when he can thereby save more than two brothers, or four half brothers or eight first cousins." (A similar quip is often attributed to J. B. S. Haldane, one of history's most quotable scientists, although Haldane never wrote it down.) The arithmetical precision of Hamilton's formula gives it an almost comical ring, and the line sounds at least a little tongue-in-cheek. But Hamilton, one of the progenitors of the theory of kin selection, took the meme seriously enough to use it in developing what is now called Hamilton's rule:

$$rB > C$$

According to the rule, genes tend to increase in frequency when the degree of genetic relatedness between a giver and a receiver of an altruistic act, multiplied by the benefit to the receiver, is greater than the reproductive cost to the giver.

The adaptive advantage of taking care of other organisms with whom we share a close genetic relationship makes intuitive sense. After all, many different branches of life have repeatedly converged on the strategy of parental commitment and sacrifice. Still, I'd bet that most people who aren't geneticists would assume that Hamilton's "world of model organisms" is exactly that: an abstract template that real living things will ignore, challenge, or modify in countless creative and surprising ways.

No matter how valid the still-controversial ideas of kin selection might turn out to be, they possess a strange beauty that I find irresistible. In particular, I confess that Hamilton's rule—with its ingenuity, urgency, and rigor—brings out in me an entirely useless feeling of love for a species that is compelled to discover a mathematical underpinning for the uses of that feeling.

Human beings are recent upstarts in the game of life, and only a few generations separate any two of us from a common ancestor. In fact, given that life on Earth preserves at least a couple of hundred legacy genes across all of creation, the genetic relatedness between any two living organisms is narrower than most of us suspect. Any enterprising geneticist should be able to calculate the degree of genetic kinship between me and a band of endangered eastern gorillas. That number, run through Hamilton's formula for sacrifice, would then compute how many gorillas (if there are enough of them remaining) a model organism like myself should feel compelled to lay down my life for, on genetic logic alone. Exact numbers for other genetically sensible sacrifices might be calculated for lemurs, leopards, lichens, or linden trees. Or, for that matter, vipers, viburnums, and violet ground beetles. Given the high percentage of genes that all eukaryotic cells share, the sacrificial thresholds might be surprisingly, uncomfortably low.

Of course, simple genetic kinship will never provide anyone's ultimate math for tying his or her fate to another's. No one ever got Hutus and Tutsis to lay down their arms and broaden their group loyalty by appealing to their shared genes—let alone all of humanity and the countless species we are exterminating. But the scope of perceived kinship does seem to grow in proportion to the threats from outside. Enemies thrown together in a hostile place, depending on each other to survive: it's a perennially popular story. If bizarre, predatory creatures erupted from underneath the miles-thick crust of ice on Saturn's moon Enceladus and headed toward Earth's storehouses of usable energy, the same species that has precipitated the loss of half of Earth's large animals since

1970 would surely sacrifice almost anything to save the last of carbon-based life.

I myself have four siblings, whom I really should call more often. But God help me, when I first came across Hamilton's rule, I could not help myself thinking, by another terrible calculus that I can't blame on natural selection, *Well, yes, lay down my life, perhaps . . . but for which two*? My knee-jerk fecklessness, though, contrasts with those countless people who, under fire, have given their lives without a moment's hesitation to save others with whom they share no close blood ties. People die willingly for spouses and brothers- and sisters-in-law, for people in their town, for their country, for strangers in an improvised clan that exists nowhere else but the mind.

Here's the astonishing thing about humans: people are always willing to lay down their lives on behalf of far fewer shared genes than any mathematics dictates. When a person scrambles up an ancient redwood that is about to be cut down, she is effectively declaring, "If you want to kill it, you'll have to kill me, too." Certainly there's a game of cultural chicken out there, a reliance on the sanctity of human life that almost every human shares. But such an action also stands ready to assert the remarkable equation: my life for this tree's. There are humans who can see cousins in creatures where other people see only otherness.

The Gospel of John puts forward a very different take on Hamilton's rule for meaningful sacrifice: "Greater love hath no man than this, that a man lay down his life for his friends." Perhaps most of the kinship we ever feel is more intuited than inherited. Perhaps genes aren't the only thing that we've been shaped to try and save. Maybe altruism evolves to recognize affinity, joint purpose, shared values. Maybe nothing elicits a sense of relatedness more deeply than feeling our dependence on other living things. A predator depends entirely on its prey; that, too, is a kind of blood bond.

Conscious creatures have an extraordinary capacity to form attachments without regard to the cost to their own individual fitness. Kinship is the ability to see my fate in theirs, even when the family resemblance is largely a leap of faith. This impulse toward expensive, even disastrous acts of identification—call it love—may be merely an epiphenomenon, a fluke spin-off of runaway kin selection misdirected into bonds with no apparent adaptive advantage, or perhaps even a net loss.

Or it may be that our capacity to will ourselves into kinship with seemingly remote creatures is in fact an exaptation: something that natural processes could never have selected for but that now (as with so many other co-opted genetic processes) presents altogether new opportunities for thriving. It seems to me that this ability to see our fortune contained in the fortune of others, a kinship based not on relatedness but on common cause, may be the one feature of self-awareness capable of saving our species from all the other potent (and potentially fatal) adaptations that evolution has endowed us with.

The ability to sacrifice self on behalf of the stability of an entire ecosystem would surely have adaptive advantage; few geneticists would disagree. But fewer would believe that natural selection alone has any possible mechanism by which to shape such a trait. That's where consciousness and fiction come in. To move from genetic kin selection to cultural selection, and from cultural selection to compassion for the more than human, a person must be driven by a calculus beyond "fitness." Hamilton's rule, applied to the messy domain of human social relations, gives way to the idea of nurture kinship, in which community tends to be strengthened and sustained when the perceived kinship between nurturer and nurtured yields more to the community than it costs the nurturer to give. Blood ties give way to proxy relations and fictive kinship—kinship grounded in shared place, shared practices, and shared narratives, both measurable and imaginary.

I am a novelist. All day long, I try to inhabit the hearts and souls of people who have never existed in the hopes that existing people might find, in these made-up lives, fictive kin who resemble their friends and provisional families in that realm of consensual fiction that we call the real world. In my fiction, kinship forms through conflict. Through the play of dramatic tension between seemingly inimical values, my characters come to recognize the keys to themselves that others hold.

Secret kinship with the Other—even with the ultimate enemy—is the lifeblood of fiction. (Surely you had to suspect, with a name like Darth Vader, a coefficient of genetic relationship hiding somewhere in the closet?) Leslie Fiedler once made the case that a great number of the canonical works of American literature have involved a plot in which a white person and a nonwhite person, thrown together in emergency, develop mutual dependence and homoerotic love. Fiction challenges the barrier between "us" and "them." It puts relations through the wringer, mangles them, and leaves the idea of family flattened but so much larger.

We're now in the middle of a family emergency that will test all family ties. Only kin, and lots of it, from every corner of creation will help us much in the terrible years to come. We will need tales of forgiveness and surprise recollection, tales in which the humans and the nonhumans each hold half a locket. Only stories will help us to rejoin *human* to *humility* to *humus*, through their shared root. (The root that we're looking for here is *dhghem*: Earth.) Kinship is the recognition of shared fate and intersecting purposes. It is the discovery that the more I give to you, the more I have. Natural selection has launched all separate organisms on a single, vast experiment, and kinship glimpses the multitudes contained in every individual organism. It knows how everything that gives deepest purpose and meaning to any life is being made and nurtured by other creatures.

Can love, in its unaccountable weirdness, hope to overcome a culture of individualism built on denying all our millions of kinships and dependencies? That is our central drama now. It's the future's one inescapable story, and we are the characters who will steer that conflict to its denouement.

To find the stories that we need, we would do well to look to the kinship of trees. Trees signal one another through the air, sharing an immune system that can stretch across miles. They trade sugars and secondary metabolites underground, through fungal intermediaries, sustaining one another even across the species barrier. But maybe such communal existence shouldn't be all that surprising. After all, everything in an ecosystem is in mutual give-and-take with everything else around it. For every act of competition out there, there are several acts of cooperation. In the Buddha's words: A tree is a wondrous thing that shelters, feeds, and protects all living things. It even offers shade to the axe-men who destroy it. Incidentally, the same man once said: The self is a house on fire. Get out while you can.

I write this on the morning after the president of the United States, seeking power in the very opposite of kinship, asked from onstage in front of an adoring crowd, "Who do you like more, the country or Hispanics?" A few days ago, his administration announced its intention to roll back the Endangered Species Act. On such a morning, I can't help but feel that we are so lost. Our capacity for salvation through sacrifice seems to be shrinking, even as catastrophe draws near.

But we make the leap in ones and twos, helping each other across like so many fictive siblings in one great adopted clan. The key is to recall how we each succeed by virtue of endless others, and that none of us will end until we all do.

Hamilton himself—he whose equation compels any life to sacrifice itself for its eight first cousins—knew this more than most of us. In an essay called "My Intended Burial and Why," he writes:

> I will leave a sum in my last will for my body to be carried to Brazil and to these forests . . . and this great Coprophanaeus beetle will bury me. They will enter, will bury, will live on my flesh; and in the shape of their children and mine, I will escape death. No worm for me nor sordid fly, I will buzz in

> the dusk like a huge bumble bee. I will be many, buzz even as a swarm of motorbikes, be borne, body by flying body out into the Brazilian wilderness beneath the stars, lofted under those beautiful and un-fused elytra which we will all hold over our backs. So finally I too will shine like a violet ground beetle under a stone.[1]

To *be many*—what more does kinship strive for? To live on in some distant branch of the spreading experiment, even in the buzzing urgency of a corpse-eating beetle: now that is how to make a common cause. *In the shape of their children and mine*—a person could find real family, out there. The math seems sound to me, and it makes a magnificent story. For to steal from that most quotable scientist J. B. S. Haldane: God Itself has an inordinate fondness for both beetles and stars.

NOTES

1. W. D. Hamilton, "My Intended Burial and Why," *Ethology Ecology and Evolution* 12, no. 2 (2000): 111–22.

PLANTS, COLLECTIVE METAPHYSICS, AND THE BIRTHRIGHT OF KINSHIP: AN INTERVIEW WITH MONICA GAGLIANO

Steve Paulson

Over the past two decades, a remarkable body of scientific literature has emerged on the intelligence and sheer virtuosity of plants, and a maverick group of scientists and philosophers now talk about "vegetal consciousness." One of the leaders in this field is Monica Gagliano, an Italian evolutionary ecologist who is a senior research fellow at the University of Sydney in Australia.

Her groundbreaking experiments with *Mimosa pudica,* known as the "sensitive plant," because it instantly closes its leaves when touched, revealed that plants learn and remember, shattering the shibboleth that only animals with brains and neurons have this capacity. Another experiment showed how peas make decisions when navigating a maze, choosing to grow toward flowing water that they can actually hear.

Gagliano has also written a highly personal memoir, *Thus Spoke the Plant,* that revealed another dimension of her life. She recounts a series of prophetic dreams, along with trips to Peru and the United States to study with shamans and indigenous healers, with whom she had experiences with ayahuasca and other psychoactive substances. Her experiences of "talking" with plants and even receiving instruction from them have stretched the boundaries of what scientists are willing to divulge.

I talked with Gagliano at Dartmouth College, where she was a visiting scholar. We spoke about the new field of plant intelligence and her efforts to reconcile the seemingly conflicting realities of modern science and indigenous ways of knowing.

Steve Paulson: Was there a particular moment or turning point when you realized you wanted to devote your professional life to plants?

Monica Gagliano: It feels more like it was decided for me. And then I had only a choice. Either you follow or you follow. So guess what? I followed.

Steve Paulson: Can you explain what happened?

Monica Gagliano: I originally trained as a marine ecologist and worked with animals, specifically fish on the Great Barrier Reef. I was in the water most of the time, so I spent a lot of time getting to know these animals, not just as my objects of research but also as subjects. And then one day, the fish taught me a big lesson: "We're not here just for your science. We are here because we have our own lives, our own stories to tell." That kind of changed a bit of everything.

Steve Paulson: Because you realized at that point that to do your research, you had to kill these fish to study them?

Monica Gagliano: Exactly. And it became very clear to me that there was no question that was important or big enough to justify me killing another being. So, yes, that kind of triggered a professional as well as personal crisis. And I had just moved from one side of Australia to take up a position in western Australia. I moved into a new home and started a new veggie patch. As I was planning the garden, planting and creating this new space, it became very clear, while I had my hands in the dirt, that I could use plants to ask the same questions. And why haven't we asked these questions with plants before? We don't want to really face the answer, whatever the answer is, especially if the answer points to the fact that plants, like anything around us, are actually alive and kicking. And they have their own story to tell. They're not just sticks in the ground.

Steve Paulson: Some of the words you use to describe the response of plants in your experiments is that they *understand* and *learn*. And you're suggesting they *remember*. We do not normally associate those functions with plants.

Monica Gagliano: For me, to use the words *memory* and *learning* feels totally appropriate. I know that some of my colleagues accuse me of anthropomorphizing. But actually, there is nothing anthropomorphic about this. These are terms that refer to certain processes. And memory and learning are not two separate processes. They are one and the same. You can't learn unless you can remember. So if a plant is ticking all the boxes and is doing what you would expect a rat or a mouse or a bee to do, then for me she's passing the test, and I don't really care whether it is a plant. The test is being passed and whoever passed the test can be said to be learning or remembering.

Steve Paulson: You are describing a level of intelligence in these plants that's pretty sophisticated. Do you have a working definition of *intelligence*?

Monica Gagliano: That's one of those touchy subjects. Intelligence really underscores decision making, learning, memory, choice. As you can imagine, all those words are also loaded. They belong in the cognitive realm. That's why I define all of this work as cognitive ecology.

Steve Paulson: This is a very controversial position among scientists. The common criticism of your views is that an organism needs a brain, or at least a nervous system, to be able to learn or remember. Of course, plants don't have neurons. Are you saying neurons are not required for intelligence?

Monica Gagliano: Science is full of assumptions and presuppositions that we don't question. But who said the brain and the neurons are essential for any form of intelligence or learning or cognition? Who decided that? And when I say neurons and brains are not required, it's not to say that they're not important. For those organisms like ourselves and many animals who do have neurons and brains, it's amazing. But if we look at the base of the animal kingdom, sponges don't have neurons. They look like plants because, when they're adults, they settle on the bottom of the ocean and pretty much just sit there forever. Yet if you look at the sponge's genome, they have the genetic code for the neural system. It's almost like, from an evolutionary perspective, they simply decided that developing a neural system was not useful. So they went a different way. Why would you invest that energy if you don't need it? You can achieve the same task in different ways.

Steve Paulson: Your critics say these are just automatic adaptive responses. This is not really learning.

Monica Gagliano: It's the job of science to revise itself. It's the job of science to be humble enough to realize that we actually make mistakes in our thinking, but we can correct that. Science grows by correcting and modifying and adjusting what we once thought was the fact. Can you as a scientist create the space for these others to express their own, in this case, "plantness," instead of expecting them to become more like you? The plant is doing this in a plant way. So can we appreciate the plantness or the animalness or the humanness of different beings?

Steve Paulson: There's an emerging field of what's called vegetal consciousness. But I think there's a popular understanding that for an organism to be conscious, it has to have an emotional life. Do you think plants can feel pain? Can they experience joy?

Monica Gagliano: It depends on what you mean by *feeling* and *joy* and all of that. It also depends on where you are expecting the plant to feel those things, if it does, and how you recognize them in a human way. I mean, plants might have more joy than we do. It's just that we don't know because we're not plants.

We have only talked about this from the scientific perspective, which is the Western view of the world. It comes straight from the Greeks, the Romans, and the medieval times. All of those dudes—because most of them were men—had certain ways of seeing the world. And we also know that this perspective has literally occupied the worldview of others.

But the worldview that I'm talking about, for example, is the Indigenous worldview. Why is that less valuable? Why is that way of knowing completely wiped under the carpet as if it doesn't have anything to say? When you actually do explore those perspectives, they require your experience. You can't just understand them by thinking about them. My own personal experience tells me that plants definitely feel many things. I don't know if they would use those words to describe joy or sadness, but they are feeling bodies. We are feeling bodies.

Steve Paulson: You've studied with shamans in Indigenous cultures, and you've taken ayahuasca and other psychoactive substances. Why did you seek out those experiences?

Monica Gagliano: I didn't. They sought me. So I just followed [*laughs*]. They just arrived in my life. You know, those are important doors that you need to open, and you either walk through or you don't. I simply decided to walk through. I had this weird series of three dreams while I was in Australia doing my normal life. By the time the third dream came, it was very clear that the people I was dreaming of were real people. They were waiting somewhere in this reality, in this world. And the next thing, I'm buying a ticket and going to Peru, and my partner at the time is looking at me like, what are you doing? [*Laughs*]. I have no idea, but I need to go. As a scientist, I find this is the most scientific approach that I've ever had. It's like there is something asking a question and calling you to meet the answer. The answer is already there and is waiting for you, if you are prepared to open the door and cross through. And I did. So I think it was a very scientific approach to go and explore.

Steve Paulson: What did you do in Peru?

Monica Gagliano: The first time I went, I found this place that was in my dream. It was just exactly the same as what I saw in my dream. It was the same man I saw in my dream, grinning in the same way as he was in my dream. So I just worked with him, trying to learn as much as I could about myself with his support.

Steve Paulson: This was a local shaman whom you identify as Don M. And there was a particular plant substance, a hallucinogen, that you took.

Monica Gagliano: I did what they call a *dieta*, which is basically a quiet, intense time in isolation that you do on your own in a little hut. You are just relating with the plant that the elder is deciding on. So, for me, the plant that I worked with wasn't by itself a psychedelic in the normal way of thinking about it. But of course, all plants are psychedelic. Even your food is psychedelic, because it changes your brain chemistry and your neurobiology all the time you eat. Sugars, almonds—all sorts of neurotransmitters are flying everywhere. So again, even the idea of what a psychedelic experience is needs to be revised because a lot of people might think that it is only about certain plants. And I find that all plants are psychedelic. I can sit in my garden. I don't have to ingest anything and I can feel very altered by that experience.

Steve Paulson: So there was a very particular regimen that you went through under the guidance of the shaman. What happened?

Monica Gagliano: I learned what I needed to learn, and then I left and took that knowing with me, which took a long time to unpack and unfold, of course. And still, sometimes little things arrive in my head and I'm like, "Oh, my God. I know where that comes from." And it took ten years to actually open up and show itself and provide the meaning that makes sense to me so that I can use that bit of information to do whatever it is that I need to do.

Steve Paulson: Did you feel like you were communicating with the plants when you were in Peru?

Monica Gagliano: I communicate with everything all the time. So not just in Peru. Not just with that plant. Once you start working with these plants in this way—and maybe this is why the *curanderos* in all sorts of traditions are so proud of those relationships—these others are living with you. So you are not just one, but you become a collective. And I guess we are always a collective. In literal terms as well. You know, we have viruses and bacteria living with us all the time. But to have also these other metaphysical aspects indicates our metaphysics is not singular. We are a collective. So, yeah, it's almost like, well, the plants are here. The ones that I have dieted with, the ones I really spent time with to build the relationship, I did as you would with any friends of yours.

Steve Paulson: It sounds like these plants were your teachers. Did they have specific information to tell you about your life and your work?

Monica Gagliano: Yeah, I mean, some of the plants tell me exactly how wrong I was in thinking about my experiments and how I should be doing them to get them to work. And I'm like, "Really?" I'm scribbling down without really understanding. Then I go in the lab and try what they say. And even then, there is a part of me that doesn't really believe it. For one experiment, the one on the Pavlovian pea, I was trying to address that question the year before with a different plant. I was using sunflowers. And while I was doing my *dieta* with a different tree back in Peru, the plant just turned up and said, "By the way, not sunflowers. Peas." And I'm like, "What?" People always think that when you have these experiences, you're supposed to understand the secrets of the universe. No, my plants are usually quite practical [*laughs*]. And they were right.

Steve Paulson: What do you do with these kinds of personal experiences? You are a scientist who's been trained to observe and study and measure the physical world. But this is an entirely different kind of reality. Can you reconcile these two different realities?

Monica Gagliano: I have the feeling now in hindsight that I was all along the target, that the plants were calling me to experience what I've experienced exactly for that purpose. It's like: can you become comfortable walking between these two worlds? They look so separate and different, but actually they are the same, but most of the time you are just only paying attention to half of it. Can you learn to pay attention to all of it all the time? I think there are some presuppositions that a scientist should just explore the consensus reality that most of us experience in more or less the same way. But I don't see why. From its earliest centuries, science has always been about experimentation and observation. And it is always personal, because it's you observing and it's you experimenting as the scientist. So I don't really have a conflict because I find this is just part of experimenting and exploring. If anything, I found that it has enriched and expanded the science I do.

Steve Paulson: You have said that you've come to think of a plant as a subject rather than an object. Can you explain what you mean?

Monica Gagliano: It's very simple, because if you objectify not just plants but even animals or other humans, you are actually allowing or giving yourself permission or justification to do whatever you want because objects don't have rights, objects don't have their own identity. They don't scream at you because they're not happy. They don't try to sue you. From that perspective, objects are something you use. In fact, even the UN definition of *biotechnology* is about uses of animals and plants, which, of course, is what GM [genetic modification] technology is about. Biotechnology uses these others as objects for whatever they want—in this case, for business and profit. For me, the question of use and the question of object are one and the same. Therefore, I cannot work with someone, whether a plant or a human or an animal, and not recognize that they have their own story. So they are subjects in themselves and they have their subjective story to tell. So I don't use them, but I collaborate with them.

Steve Paulson: What does it mean to think of a plant or to interact with a plant as a subject? What changes when you look through that lens?

Monica Gagliano: There is an honoring of the life that is there. By objectifying, you take away the life.

Steve Paulson: It sounds like you're saying the plant has some sense of being, or, to use different language, a soul. And maybe that extends even to inanimate objects, like rocks and rivers. Do you see everything around you as alive? I mean, in an animistic sense, is there beingness within everything in the natural world?

Monica Gagliano: It shouldn't really be that controversial, but we're still discussing it. And what you're describing is exactly how I feel about the entire question of consciousness. For me, consciousness doesn't arise from matter, but it is the substrate from which different forms of matter arise. And from that perspective, there is no conflict to think that plants would be conscious, because consciousness is to me what we call essence. That is, you know, life. And life takes many forms. Some of them are biological and some are not.

I'd rather assume that consciousness is everywhere and then we can refine later, rather than the opposite, because if we get this wrong, that would be really quite serious. Wouldn't it be safer to just allow for it rather than to decide a priori that it can't be?

Steve Paulson: You are challenging centuries of Enlightenment thinking and presenting a different way of knowing the world than the scientific materialism of the vast majority of scientists.

Monica Gagliano: Yeah [*chuckles*]. Is that a problem? Are they going to burn me at the stake? They probably would if they could. But it's not kosher in our society to burn people at the stake because they think differently.

Steve Paulson: More and more people these days are talking about kinship with the more-than-human world. Does this idea of kinship resonate with you?

Monica Gagliano: Yeah, I like the word. *Kinship* contains this idea of family. It extends the concept of family beyond my blood sister or blood brother. And then you extend it to the nonhuman and even to place. You know, there are places where I just feel love for. So I think kinship and love are very closely connected. And it's important at this time in particular because we care for things we love. We definitely don't destroy them because we don't want to see them die. So kinship can be a very potent medicine for this time.

Steve Paulson: So you're saying kinship is not just an idea or a theory. There's a practice that's involved, there are implications for how we should live.

Monica Gagliano: And here at Dartmouth I was talking with colleagues from the environmental humanities, which is a relatively new academic area. And they were telling me how the new students are flocking to these courses on environmental humanities because it's one of the few places that brings together the best of the humanities and the best of environmental science. The students realize that we can't separate those two. You know, the humanities allow you to talk about, for example, love or empathy. And science allows you to work out which method would work to care about this. So when you put the two together, you're really empowering people to think and act in a way that is based on kinship.

Steve Paulson: As I listen to you, I get the sense that you are giving us a world that is magical and enchanted. If you could spread that vision to all of us, how would the world change?

Monica Gagliano: First of all, I'm not giving you anything.

Steve Paulson: I know, I don't mean to turn you into a guru, but if a lot of people saw the world the way you're describing it, what would it look like?

Monica Gagliano: It would be a very different place. The thing is, we all see it exactly as I'm describing it when we are kids. So whatever practice works for you. Maybe you are an ultramarathon runner and that's what gets you in that soft spot where you are feeling connected to the world in a very deep and profound and kind of mystical way. Or you're doing yoga, or meditation, or when you are reading—who knows what works for you? But whatever it is you do, when you find yourself arriving in that place, expand that space and take it everywhere you go. Then you are changing the world because you're creating a vision that is very different from the one of disconnection and lack of care and lack of love that we are experiencing at the moment because we are responsible for dreaming it that way. So this is almost like it's part of our heritage, because as kids we all come as totally magical beings in a magical world. And then we kind of get lost because we're trained to think,

This is not how we do it. But we can also train ourselves and others in a different way. We can remind each other that this place is actually pretty amazing. If you really stop and think about what this place is and how incredible it is that you can breathe, that is a pretty psychedelic experience in itself. And then you will start seeing everything and everyone in a different way. You know, it's not difficult because it's our birthright.

SKYWOMAN'S GARDEN

Rowen White

Winter is the sacred season of story and dream. For my ancestral people, the Haudenosaunee, Midwinter is a time for sacred beginnings, when the infinity loop of time resets and the sacred fire must be put out and restarted with prayers and blessed with new intention. It is the beginning of our ceremonial cycle, and it reconnects us to the place of our origins, where our Original Ancestor fell from the sky, the sacred smoke hole in the sky known to many as the Pleiades star constellation.

When the "Seven Sisters" are directly overhead, it is our cue to spend days in prayer and healing. It is when we see ourselves as seeds in the moist and dark earth, allowing our seed coats to imbibe the Story Water of Life and begin to crack open in anticipation of the coming spring. We must seed our prayers well in advance of the sprouting of the Earth, to nourish and feed the keepers of Life. It is an honor and privilege to tend this exquisite tension of Life wanting to bless us with abundance.

Midwinter is the beginning of our agricultural cycle, the time when the seeds lie dormant, just like the seeds in the dark of the soil before the spring sprouting. Midwinter is the time of seed dreaming; for some this comes in the form of gathering the stack of seed catalogs that arrives daily in our mailboxes; for some it is digging into their treasure box or shelf of saved seeds from previous seasons, tiny little gems of seeds that hold such embodied potential. For us, as Mohawk people, it is the time we renew our relationship with our original foods and sow another season of connection in the grand dynamic coevolutionary dance that we have with our sacred foods and ancestral homelands.

Years ago, as I accepted sacred bundles of ancestral seeds of my Haudenosaunee ancestors, I began a decades-long, unconventional, rite-of-passage ceremony in which the plants and the Earth helped grow me into a deeper understanding of what it means to be a Mohawk woman in my full capacity. I was told by wise elders that there is a cartography of stories that reside in the landscape that help us as Mohawk people remember who we are, where we come from, and how to live in a way that honors our responsibility to be a descendant of good mind and a bright future ancestor. As we traverse the seasons, we walk across a memory landscape of stories that help us remember our place in the mycelial network of kinship that is all around us.

It was Original Woman, who came here to this layer of existence, clutching a handful of seeds, and proceeded to sing the World awake, sowing her seeds into the Earth that was on the back of a Great Turtle. She came tumbling down from the Skyworld into a watery abyss, and all our relations here conspired to make a suitable home for her and all of us, her descendants. This beautiful, layered creation story, which began so long ago, continues to unfurl and come alive in every moment, as our original life sustainers emerge from the soil in our gardens and sing a song that helps us remember our Original Agreements to take care of one another.

Dancing in the direction that the sun goes, First Woman put into place the cycles of continuous creation, continuous birth inside the Earth and all around us. Skywoman gave birth to a beautiful daughter soon after arriving here in this new world, and in time she too bore the two twins who would continue to help shape the place we call home. During the birth of these Original Twins, Skywoman's daughter passed in labor, and in her dying breaths, she claimed that the foods needed to sustain her descendants until the end of time would sprout from her body in her earthen grave. True to her word, from her breasts sprouted the Corn, who would grow in diversity to feed both villages and empires. Twining bean vines emerged from her hands to offer long and slender pods filled with nourishing seeds, and raucous squash tendrils grew from her belly button to produce sweet and sustaining pumpkins and squashes to feed people well into the deep of winter. Sunflowers and Original Potatoes grew from her legs, Tobacco for our prayers from her head, and the life-affirming first fruits of the Strawberry emerged from her heart.

Seeing that these original foods grew from the body of one of our first ancestors, we understand that we are lineal descendants of these foods that nourish us. We understand that these seeds are our relatives and we are bound to them in a mutually beneficial relationship with them since the dawning of time. These agreements to care for them are held in the earth of our bodies as well as the soil into which we plant them. We plant these seeds into good earth, season after season, to renew our commitment to tending a relational kincentric way of nourishing our communities and families, just as countless ancestors have done in our ancestral homelands of Turtle Island.

We plant our seeds each season in a continuation of prayer and ceremony of remembrance of these threads of kinship and lineage that have been given to us forever, with the promise that we would have good food to eat. As an Indigenous woman returning to my traditional lifeways after many generations of disruption from the unspeakable cultural erasure and violence of genocide, acculturation, and assimilation of colonialism, I have to be a patient listener to the memory of the old ones (human and not), who left achingly beautiful stories and songs in our blood, in our bones, and in the landscape all around us, just waiting for the monsoons of time to rain down and coax them into sprouting once again in our modern lives.

It is this same Original Woman who upon her parting from Earth returned to the Sky in the form of Grandmother Moon, who continues to look over us, guiding the cycles of life, fertility, death, and rebirth with her lunar rhythms. Her beautiful luminous changing face pulls at the salty oceans that blanket a large percentage of our planet to create the tides. It is this same gravitational force that moves the smaller oceans that reside inside our own bodies, and the tiny pools of water that exist within everything that is alive; the minuscule oceans that sit inside the heart of every seed, the sapling trees, the supple bodies of deer and fox in the forest.

Naturally, our Grandmother Moon still watches over the cycles of our plant kin, as they move from tiny seeds into sprouts and around the seasons into buds, blossoms, fruit, and seeds once again. Our ceremonial cycles celebrate this kinship, this connection of our Original Ancestors, and our cosmogenealogy of relationships. From Midwinter where we bless the courage of the spirit of the seeds who dream and sleep, to the opening of the Maple Trees who are the harbingers that winter will soon turn to spring, to the Thunderers who send their lightning bolts of energy to awaken the soils, to the vibrant clatter of the peach stone game which helps us humans awaken the seeds; each turning of the season brings ceremonies, songs, stories, and a rhythm of everyday rituals that helps us remember that we all play our part in keeping the cycles of creation alive and animated.

This summer, during the raging wildfire season that turned the sun bloodred in the ash-laden sky, I instinctively anchored into tending what was most important: tending the health and safety of my relatives in immensely transformative times. To calm my limbic system, and to find the headwaters of my breath in a time of great ecological distress, I made myself small and quiet, crawling into the tangle of the patch of Corn, Beans, and Squash in my garden, a sanctuary where I have apprenticed myself to the food plants that have become my grandmas and wise elders over the decades that I have had the honor of tending them. There in the garden, close to the living, breathing, pulsating Earth, I found the headwaters of my breath again, tuning my senses to notice the million tiny miracles of expressions of life all around me, the countless ways Mother Earth sings her songs: tracks of little paws in the dry, red-clay dust; tiny little fuzzy, nascent watermelon fruits clinging delicately to vines; the sensuous ways beans were dancing counterclockwise around the corn stalks. Bean leaves sticking to my shirt like Velcro. The fine little hairs around the edges of each leaf and stem. The cool exhale of plant stomata breathing. Beetles making funny little clicking noises as they meander. The crackling, static song of the hummingbird perched in the nearby pine. We shared the same breath, and I felt in the Earth of my body that the same essence that animates these beings all around me courses inside of me, connecting us in kinship.

There I was, sitting close to my beloved Mother Earth—Ionkhi'nisténha ohwéntsia, in our Mohawk language—in all her aliveness underneath the foliage of my Three Sisters garden, planted with seeds entrusted to me by my elders, some of whom are now ancestors. I had this profoundly familiar visceral remembrance that this too was Skywoman's garden, for as our elders tell us in our stories and ceremonies, our creation stories never ended. These timeless stories continue to unfurl each and every season as we renew our ceremonial bundles and plant the seeds and sing the very same songs that Skywoman and her daughter sang when they first came to this place we know as Turtle Island. Here in my own garden, I was once again woven into these cosmologies, as I tended these tiny sprouts, which in our mutual dance of reciprocity and care has become corn, still springing from the bosom of our Mother Earth; basketfuls of beans gifted to us by her benevolent extended hands; voluptuous squash fruits that spiral from the umbilicus of the Earth. These relationships, these interwoven stories, still reside in the very ground beneath my feet and the Earth of my own body, her generosity springing forth with every handful of seeds that she offers for our nourishment as we do our best to make our lives love poems and to ensure that the ancestral bundle of seeds and stories that we've been handed is more beautiful and magnificent as we hand it down to the next generations. It is in this earth-in-hand prayer in my garden that I rehydrate those agreements, those promises, those reciprocal agreements that link me through the bloodlines through countless generations of women who came before me who kept the seeds alive in the face of unspeakable atrocities and adversities because they knew that young women just like me would one day apprentice themselves to the very Divine life cycles that have been unfurling since Original Woman's first dance on Turtle Island.

I sing a whisper of a song to give courage to all the future revolutions and epiphanies that will sprout from moments like these, as we return to the deeply relational ways of our ancestors once again. May we remember that we are woven into a beautiful tapestry of kinship of place, as we push against the modern narrative of disconnection that tells us we simply live on the Earth instead of inside of a beautiful storied relational landscape. May we lean into this practice of reverent curiosity as we enliven the stories that lay dormant in the land, allowing us to sing countless praise songs to this most exquisite Earth family that has never forgotten us. May we raise the next generation of humans who love the Earth and Moon as Grandmothers and honor the plants and seeds as relatives. Those of us who love the Earth as Mother; we are many.

RUN SALMON RUN

Christopher McLeod

It was the third day of Run4Salmon and Chief Caleen Sisk of the Winnemem Wintu was on a motorboat, heading from the salty waters of San Francisco Bay to the fresh water of the Sacramento River. A hundred pilgrims had walked for two days through the Bay Delta—Ohlone territory—following the route of migrating winter-run Chinook salmon, as we proceeded on a two-week ceremonial journey. Now a dozen of us were on the water, speeding toward historical spawning grounds in the cold rivers on the flanks of Mt. Shasta, three hundred miles to the north—if only the salmon could get past towering Shasta Dam.

Gazing down into the water, trying to be a prayerful participant, I imagined that I was a salmon. I tried to see their world. It was murky and warm—and getting scarier every day. Hidden among the high-tech windmills by the side of the river were dozens of fracking wells. New natural-gas pipes were being laid down across the river as we motored by, accompanied by five state water officials who were listening to Caleen's insights while seeing the sad state of the river for themselves, some of them for the first time.

"The mountain called for this ceremony and it's calling the salmon home," said Caleen. "That's the importance and power of this run for salmon."

Caleen pointed out how the straight, channelized river—with cars whizzing by on each side, on top of artificial levees—provides no place for adult salmon swimming upstream to rest in the shade, or for little fish headed to the ocean to feed in what used to be seasonally flooded tule wetlands.

We reached the Delta Cross Channel, an artificial canal almost as wide as the main river. Engineers and bulldozers built the canal in 1951 to move Sacramento River water toward giant pumps that push it into aqueducts leading to farms and cities far to the south. It's water from Mt. Shasta that feeds the world, and makes a handful of corporate farmers rich—and they wanted more of it.

The captain cut the motor. We drifted down the middle of the main river and suddenly were pulled toward the eastern side and straight into the Delta Cross Channel. The pumps must have been on.

"Imagine being a small, young salmon, swimming down the river," said Caleen. "You feel the current going this way, and you follow that flow—you're thinking, *The ocean must be this way*!" She nodded toward the east and shook her head. "They get sucked into the pumps, and that's the end."

Caleen calls salmon "the magical fish." The juvenile fingerlings that head down from their home river toward the ocean are three-inch-long freshwater fish. When they get to the Bay Delta—if they can avoid the pumps—they begin to transform themselves into saltwater fish and eventually grow to between thirty and one hundred pounds. For the following four to seven years, they mingle with salmon from other rivers, traveling far and wide, and feeding everyone in the sea—orcas, seals, sea lions—as well as people on land. Then, answering a mysterious call, they head back home, readapting to fresh water, starting the long upward migration against the flow of the rivers, seeking their birth streams in the high country, sensing the Earth's magnetic field, and "smelling" their original waters.

"The fish have to climb the mountain," says Caleen.

Along the way the adult salmon continue to feed eagles, otters, bears, wolves, and people. Once home in the cold, oxygen-rich water where they started, they spawn and die within a few days, giving life to the next generation and fertilizing rivers and forests.

As Caleen says, "They never stop giving." And as Run4Salmon participants learn, the journey is not only difficult because it is uphill all the way—there are new obstacles every day.

Sacramento River winter-run Chinook salmon were listed as endangered in 1994. That means they are likely to go extinct without real, substantive changes in their habitat along the West Coast of the United States. In waters warmed by climate change, pollution from farms, factories, mines, logging, and cities continues unabated. Flood control measures have altered the meandering course of once free-flowing rivers. And when the salmon approach the mountains where Caleen's people honored and ate them for millennia, they now run into the 602-foot-tall concrete impasse that is Shasta Dam. There is no bypass, no way around. As we motored up the river, the federal Bureau of Reclamation had revived its efforts to make the dam 18.5 feet higher.

The Winnemem creation story tells of a time when all of the spirit beings were choosing physical form and determining their role in taking care of Earth. All of the beings—eagle, bear, hummingbird, salmon, fire, and water—were coming out of a sacred spring on Mt. Shasta. All had chosen their jobs but one being, who seemed lost and without clear purpose—the human. Creator called fire, water, and salmon back to assist the human. "You're going to need to help this one," said Creator. Salmon spoke up, and offered to give humans their voice, figuring humans would do better if they could communicate. The Winnemem people agreed at that sacred moment to always speak for salmon.

Indigenous communities from California to Alaska to Kamchatka share this appreciation and respect for salmon, as humans continue to recognize the seasonal gift that rivers bring, the cherished food that has sustained people forever. This universal sense of reciprocity has evoked songs, dances, and ceremonies as dozens of salmon-dependent cultures celebrate their relationship with a magical fish, enacting a gift-exchange between species.

At one ceremony, I overheard a woman shaman from Kamchatka, Russia, say to Caleen, "When the salmon are running they fill the peninsula with the feeling of love."

Along with her small, determined Winnemem Wintu Tribe, Caleen has never stopped fighting for the salmon—and she needs an army of allies. She's found strong support in the community that has formed over four years of Run4Salmon, an extended direct action that balances spirituality and politics, and treats the two as indistinguishable. Over the course of the two-week pilgrimage, people come and people go. Some participate for a day, some for fourteen. Along the way, these water protectors and prayer warriors make superb use of social media. It is educational storytelling in motion over three hundred miles, for distant relatives, in the name of kin. For many, it is a life-changing experience: protecting sacred land and water under the leadership of local Indigenous people. By the end, everyone is elated and exhausted. Like the salmon, we all swim in troubled waters.

When the boat docked in Sacramento, we were met by a dozen Miwok singers who welcomed Caleen and the Winnemem as the sun set. The following day, fifty red-shirted activists stood outside the domed state capitol chanting and giving media interviews. Six teenage girls from the New Village School in Sausalito held up individual letters spelling out "No Dam Raise." We sang our way into Governor Gavin Newsom's office, and Caleen delivered a letter demanding that he sign SB1, a bill passed by the California legislature committing the state to fight the Trump administration's efforts to weaken the Endangered Species Act, a direct threat to salmon recovery. A couple of days earlier, Newsom had unexpectedly announced he would veto SB1, honoring his ties to corporate water and agribusiness interests that provided him with $637,398 in campaign contributions. A chief from the Amazon happened to be in the capitol to oppose carbon-trading schemes that threaten Indigenous land rights in his homeland. He joined Caleen for a photo op, and then invited her into his meeting with Newsom's staff. Sacred site guardians look out for one another and are especially sensitive to the irony of politicians who are willing to meet with foreign Indigenous leaders while ignoring those who are their constituents.

Continuing north out of Sacramento on bicycles, we saw almond and walnut monoculture, widespread fracking, and sluggish waters in surreal, straight-as-an-arrow concrete aqueducts. We paused at a giant oak tree in Tehama County, where gold-rush settlers lynched Maidu and Wintu people and stole their lands. There was a cryptic sign that contained no information but a name: "The Witness Tree." A prayer ceremony was held there, tobacco and water sprinkled on the roots.

When it was my turn to run in the relay procession, I carried the eagle-feathered salmon staff through an oak grove along the Sacramento River, with Mount Shasta straight ahead on the northern horizon. I felt fully grounded, a link in an unbroken but fragile chain, part of a vibrant community reaching out to ancestors, salmon kin, and future generations.

When we reached Shasta Dam, seventy-five people flooded the visitor center to protest the story told by the Bureau of Reclamation: a glorification of empire and progress and the taming of wild nature. The Winnemem story was entirely absent, no mention of how the dam submerged eighty-two historic village sites, numerous burials and sacred places, and more than a hundred allotments inhabited by Winnemem families when twenty-six miles of the McCloud River was turned into a giant reservoir.

The next morning, I climbed into the front of a two-person kayak with a young, strong friend from Chicago, the writer Gavin Van Horn, in the back. Amid a fleet of twenty kayaks and canoes, we began to paddle across Shasta Reservoir (not "Shasta Lake" as the government would have you believe). It was a beautiful blue-sky day as we glided over the drowned Pit River, over submerged copper smelters, toxic-waste dumps, and towns that all disappeared when Shasta Dam was completed in 1945.

When we turned north up the McCloud River canyon, the culmination of twelve days of travel by foot, bicycle, boat, and now kayak—plus one thousand feet in elevation gain—we had finally entered the heart of Winnemem Wintu territory. Making camp that first night beneath a tall mountain known as Gray Rocks, I could only imagine the villages and the history beneath the rippling water. It was right below us that biologist Livingston Stone proposed building a fish hatchery in the 1870s, only to be confronted by Winnemem men who staged a war dance to defend their river. Under this very water was the village where the renowned Wintu healer Florence Jones grew up and started her initiatory forty-mile solo pilgrimage to Mt. Shasta when she was just ten years old. Years later when the dam was finished and the valley began to flood, Florence helped move 183 burials—including her parents and baby—disinterring them from family cemeteries beneath Gray Rocks and reburying them in a new government-controlled cemetery near the dam.

With no way to get around the massive dam, salmon threw themselves at the concrete wall while homeless Winnemem stood and watched. There would be no salmon in the waters above the dam for many decades. The Winnemem come back year after year after year. The salmon and the Winnemem continue their struggle to give birth back in their traditional home. Caleen will never forget the debt her people owe the salmon. A gift creates an obligation. It's how it works in a family and how it works with nonhuman kin.

Paddling up the canyon was sobering and sad as we passed Dekkas Rock, where a stone feature known as "The Guardian" will go under if Shasta Dam is raised. I imagined beneath me the flat bench where elders used to meet in councils, to discuss acorn gathering and village conflicts, and where "Big Time" marriage ceremonies were held as the people feasted on salmon. Kayakers around me started singing the anthem of the day, "We're gonna stand up for the water, and for the sacred land we honor," repeating the mantra over and over as we stroked in rhythm. I closed my eyes and conjured the sounds and images of roundhouse dancers in the flooded village below performing the World Renewal ceremony to ward off earthquakes and floods and pray for abundant acorn and salmon harvests in the years ahead.

As I looked at two young Winnemem women paddling a wooden dugout canoe, leading us all up their drowned river, it was painful to imagine the feelings their Winnemem ancestors must have had as their homes were bulldozed, allotments lost, and sacred sites drowned.

But the women were smiling, and the people were singing. Our calm liquid pathway was mesmerizing, its velvet-smooth surface connected to the waters that bubble up from the sacred spring above us on Mt. Shasta. That life-giving spring is the origin place of the Winnemem people, where they made the commitment to always speak for salmon. Sing to the water, call salmon home.

We saw three bald eagles that day as we collectively prayed that the salmon Livingston Stone exported around the world more than a hundred years earlier—some of whose descendants are thriving today in New Zealand's Rakaia River—will soon make their way up a new swimway around the dam and home to what's left of the McCloud River.

And one day, of course, the dam will no longer stand. May the salmon survive until that time.

The roundhouse captain and singer Gary Thomas led three Winnemem dancers down through tan oak trees, their orange flicker-feather headbands radiant amid backlit green leaves.

They walked slowly along a ridge, twenty feet above the crowd that had assembled below in a circle around the fire. Our journey was ending.

The three barefoot dancers swayed like salmon tails, whirled and bobbed like birds. Caleen came over to a couple of Native Californian men I was standing with and asked us to move over and stand behind Gary as he sang. She nodded down toward her moccasined feet, encouraging us to stomp ours in rhythm with the drum.

"This is how we call the salmon," she said to one of the elders from Hoopa, stepping to the drumbeat and waving a blue scarf that flowed like water in motion.

As we touched the earth for the next half hour, it turned out to be the perfect place to snap some photos of Gary and the dancers, who finished each short song with a cry as they circled a few feet away.

Figure 1. Gary Thomas and Winnemem dancers. Photograph by Christopher McLeod.

The ceremony drew to a close with the whole group of a hundred people moving in a circle around the fire. The dancers exited and walked back up through the forest. In the stillness that followed, Caleen pointed at me and filmmaker Will Doolittle and signaled that we should come over to the fire. She said, “Bring your cameras.”

Caleen blew tobacco smoke over Will, placed her hand on his heart, and talked to him in a soft voice. She blew smoke over my face and hands. Then she looked down and nodded toward my camera, which I lifted so she could blow smoke on the lens and then over the body. The dance captain, Rick Wilson, smudged us from behind with sunflower root. The world seemed silent and still as warm sunlight showered down from overhead. It was a blessing, a perfect moment.

Caleen put her hand on my heart and said: “Thank you for telling our story. We need you to do it with this camera and with your words. Always tell it true.”

She looked into my eyes. “Always share our story in a good way. Even if they don’t want to listen, let’s keep telling the story, and let’s keep speaking for salmon.”

ONE SATURDAY MORNING

Sean Hill

It was spring in Carmel. There were birds. It always
feels like spring in Carmel. It still felt like deep winter
back home in Fairbanks. There were Bushtits and Band-tailed
Pigeons, Acorn Woodpeckers, Western Scrub-Jays
and a Brown Creeper, Black Phoebes, Spotted and California
Towhees, a Townsend's Warbler, and others; a gull floated
high overhead. I stood on the patio in the least layers
I'd worn in months. I held you, legs wrapped by my right arm,
between my wrist and elbow your seat. You'd only recently begun
pointing out the living room window at trucks and cars
moving along
the hardpack snow on the road back home. There were birds—
a whole new community for you—and the American Crow
on the wire was the largest, and so I thought the easiest
for your young eyes to perceive, so I pointed up at it, and you
pointed with me with what was surely my hand when
I was your age, but your eyes didn't follow our fingers.

They looked to me and out and inward like when we
listened to music in the kitchen back home in Fairbanks
with our waving hands in the air, but on the patio there
was no music save birdsong. I pulled my hand back
and jabbed it at the crow, and we're doing an old disco move.
I leaned back with you in my arm and watched your eyes
roll down in your head to stay level to the horizon.
After a couple of tries the crow cawed and ruffled its feathers
at us, and you discovered it. Oh, your emphatic finger
showing me
that there was a crow on the wire—acknowledging that bird
and then others. You revealed not what I knew—that there
were birds one Saturday morning in Carmel—but made me feel
again how the world opens before a curious body—our bodies.

KINSHIP BY THE SEA

Tim Ingold

There is no better place to observe human kinship than on a sandy beach in summertime. You can watch it here, in real time, as it is said and done, in the myriad ways in which little groups of people of all ages, from the tiniest infants to wizened elders, gather in knots, encircling themselves with windbreaks or parasols, or simply by drawing a line with a finger through the sand. Having arranged their towels, rugs, and portable furniture within the circle, they set about the serious business of enjoyment. This involves a good deal of sometimes raucous vocal communication, in which words of endearment, astonishment, and command are mixed with shouts and squeals, laughter and tears, all against the background of screaming seagulls, barking dogs, and the reassuring wash of the waves. And as the many layers of clothing worn in ordinary life are stripped off, to be replaced by swimwear—or for the youngest, nothing at all save a sunhat—there's a certain intimacy of bodily contact, for example in the mutual anointment of lotions and sunscreen and the rough-and-tumble of children's games. But once having established their arrangement and shared their provisions, the people scatter every which way, to comb the beach, peer into pools, or jostle with the waves. The knot unravels, and all we see are heads bobbing in the ocean, figures jumping at the water's edge, and—here and there—builders, armed with bucket and spade, raising monuments of sand. The scene is one of apparent chaos, yet eventually, barring disaster, all these bodies will find their ways back to base. Something draws them in and ties them together again. That something is kinship.

What is this mysterious force that seems to tie people in knots almost despite themselves? Observing the scene, you have the impression that their cohesion is not a matter of choice. For each and every participant, the others are simply *there*. Older ones have always been there, until eventually they pass away. Younger ones have arrived, not from outside but born into the midst of everyone else. Yet thrown together in life, they are fated to muddle along as best they can. Physically, of course, individuals come and go, and the knots that actually form on the beach can be of the most varied composition. But even with those who, for whatever reason, are physically absent, their presence lingers. They are there in conversation, and in memory. For example, the two brothers, aged ten and twelve, whom we now observe rollicking on the beach remember their dad, who left their mother some years ago and sees them only rarely now. They have to put up with a stepfather who arrived through no choice of their own, and a half sister, born just three years ago. But they dearly love their mum, and their little sister too, and would do anything to protect them. Their real rock, however, is granny—their mother's mother—who never scolds and is a source of endless kindness, although she is still grieving from the recent loss of her husband, the children's grandpa. About their uncle, their mother's brother, still a bachelor, the boys feel more ambivalent. He has come along for the day—he always does—but they don't quite know what to make of him, or he of them. He's just a relative, they say. Sometimes he gives them presents, and their mother instructs them to write thank-you letters.

This example, though entirely imaginary, suffices to illustrate the messiness of human kinship, with its complex and often unstable mix of passions, but grounded in a sense of presence from which it is no more possible to escape than from one's own existence. The boys' mother and their uncle once grew up together as brother and sister. They may have alternately adored and detested each other, but the memories from that time, perhaps including seaside holidays, remain seared in their hearts, profoundly affecting their sensibilities and the ways they act and respond toward others. In another twenty years or so, the two brothers—whom today we see arguing and fighting at one moment and dissolving into laughter the next—will likely have gone their separate ways, perhaps living in different places with families of their own. They will become distant. This distance, however, will never erase the closeness of their common childhood. So much is shared by siblings growing up together—not just days out by the sea but also the attention of parents, the environment of the home, the commensalism of mealtimes, even shared infections and the immunities that build up as a result—that their family resemblance is evident to all except perhaps themselves. The tragedy of kinship is that similarity repels and that closeness inevitably breeds division. It is the brothers' very intimacy, in childhood, that will eventually drive them apart. Yet in that very repulsion lies kinship's hope, that siblings, having parted company, will eventually bind themselves in other knots.[1] Repelled by similarity, lives are brought together, and find their continuation, in difference. Yet no one lives forever, and the loss of a kinsperson, however distant he or she may have grown in life, is always a cause for grief, because it is a loss of part of oneself as well.

Returning to that little gathering on the beach, however, there is one participant whom I have so far failed to mention—of course, the family dog. Loved by all, she is a constant companion: wherever the others go, she goes too. In her own way, she is as vocal as the others. They talk to her, and she responds. She always looks forward to these days on the beach, and not just because they offer a welcome respite from the relative confinement of the home. If human families are inclined to be wary of strangers, their dogs are far more sociable and never miss an opportunity to meet and cavort with others of their kind. Although the general hubbub of these canine encounters can initially induce a degree of consternation among the dogs' human handlers, who are keen in public to assert their control, it can also spark friendly banter between people who would otherwise pass each other by without a glance. Yet after every encounter, dogs are called away by their respective handlers to where they belong, in the human fold. Indeed, the dog would seem as caught up as everyone else in the traffic of affection and disaffection that we see play out on the beach. In the family, the dog and possibly her offspring share with human generations the experience of growing older together. Kinship is produced and reproduced in this experience. Does this, then, make the dog a kinsperson? Well, yes, many would say, save for one thing: he or she lacks human progenitors. At the end of the day, however affectionate the bond between them may be, dogs are descended from dogs and humans from humans. And that, it is said, rules out any possibility of real kinship.

This invocation of descent at once introduces another logic of kinship that is very much at odds with our observations up to now. Not everyone subscribes to it; indeed, if we were to compare ideas of kinship from people around the world, it may turn out to be more the exception than the rule. It is a logic, however, that holds particular sway in Western societies. These are societies in the grip of a way of thinking—nominally scientific, although not all scientists would adhere to it—that seeks truth in universals, in laws that apply to things regardless of time, place, and context. Applied to living things, this is to claim that each is determined, in its essential nature, independently and in advance of its life in the world, by a complex of attributes bestowed on it at the very moment of conception. In a word, these attributes are said to be *inherited.* Like any other kind of inheritance, such as of property, they are passed on ready-made. Life, then, is given over to their expression of fulfillment, but not to their production. Nowadays, following the popularization of microbiology, it has become common to refer to these heritable attributes as genes, even to say that they are in a person's DNA. The measure of kinship, according to this logic, is to be found in degrees of genetic connection or relatedness, given at the start, rather than in any relationships formed in early or later life. The more closely related two individuals are, the more inherited attributes they share in common. The web of kinship, then, is not woven from the threads of life, as they are gathered in the kinds of knots we see being tied on the beach. It is formed of the connections between procreative events, each of which marks the point of inauguration of a new life cycle.

This is what is meant when people say that *real* kinship is biological. Their understanding of biology has nothing to do with the life of organisms or its study. It excludes all those acts through which parents nourish and care for their children. It excludes the sharing of infection and the buildup of immunity. It even excludes the contribution that a mother makes to the growth of her child during pregnancy. Instead, biology has come to mean a program—a *bio-logos*—allegedly preinstalled in every living being and orchestrating its growth and behavior from inside. Everything else is taken to be but an expression of this genetic *logos*. Naturally, there are situations in which a direct genetic connection cannot be traced. The two brothers introduced earlier, for example, have no such connection to the man who fathered their half sister. Although he is very much a part of their present lives, genetically he is a stranger. This, however, is taken as the exception that proves the rule: the boys' real, biological father is another man altogether with whom they now have little contact. With their mother, of course, it is different. She has loved each of her sons from long before birth, ever since the first quickening in her womb told her that a child was on the way. This love owes nothing to the coincidental fact of her genetic contribution to the embryo and everything to the intimacy with which the new, emerging life is folded into the compass of her being. Yet the calculus of kinship as genetic connection has no place for love, care, nourishment, or any of the other affections that allow lives to flourish. These, it is supposed, are purely derivative, the expressions of evolved instincts that owe their establishment to impersonal forces of selection. It is the genes that make us who we are.

By and large, on the matter of kinship, mainstream bioscience has marched to the same tune as popular belief, each feeding on the other in a spiral of mutual reinforcement. This is true of the imagination of kinship not only within species, such as our own, but also between one species and others. Among individuals of the same species, the preferential treatment of kin is allegedly encoded in the language of inherited genes that, competing for a stake in future generations, gain an advantage from causing their hosts to behave in ways that improve the life chances of fellow carriers. The closer the connection between any two individuals, the greater the proportion of genes they will have in common, and so the higher the payoff. Although it might work for insects, applying this theory to creatures as complex as humans leads to conclusions that can be described only as farcical. Are we seriously expected to believe that if the children ran into trouble while bathing in the sea, the boys would instinctively help each other before concerning themselves with their more distantly related half sister? Not only does the theory offer a grotesque caricature of relations that, in real life, are haunted by multiple and often conflicting feelings; its reasoning is entirely circular. For taken on their own, genes do not encode for anything; there is no "reading" of an organism's genes that is not part of the process of its development in a richly structured and varied environment. For human beings, as for creatures of many other kinds, this environment includes others whose presence and activity helps shape their developing dispositions and sentiments. Kinship, as we have seen from our observations on the beach, is fashioned in the shared experience of generations of humans and nonhumans growing older together. To suppose that this play merely follows a script that is already written in the language of the genes is to put the cart before the horse.

In modern evolutionary theory, this circularity is endemic, and it afflicts the way we tend to think of relations between species as much as of relations within them. Are we innately disposed to favor creatures to which we have the closest genetic connection?

Among extant creatures, these are the great apes, particularly chimpanzees. We might like to think of chimpanzees as long-lost relatives, but our human record of helping them is not great. On the contrary, we have driven them close to extermination. Most of us have seen them only in pictures or, if alive, caged in a zoo. Nevertheless, when the primatologist Jane Goodall, after years of habituating a band of forest-dwelling chimpanzees to her presence, eventually shook old Graybeard by the hand, the popular press billed it as "the handshake that spanned five million years." *How many million years,* you might wonder, as you watch the children playing on the beach, *do they span every time they caress the family dog?* For them, of course, the question is irrelevant, because the phylogenetic connection between humans and dogs has absolutely no bearing on the feeling they have for the animal as a lifelong and faithful companion. More generally, an understanding of our kinship with other species on the basis of genetic proximity, or degrees of relatedness, is bought at the expense of cutting every being out from the matrix of relationships in which it lives and grows, and treating each as an isolate, attributed to its class on the basis of covert attributes encoded in advance of its entry into any real-world environment. This is not the way to restore human beings to the continuum of the living world. And the problem lies in the logic of descent and inheritance.

Dogs can be descended only from dogs, we say, and humans only from humans. But not everyone would agree. Many people around the world number, among their ancestors, mountains and creeks, birds and bears, and a host of mythical creatures. They would understand the ancestors, however, not as a source or conduit for inherited attributes, passed down the line from generation to generation, but as beings that came into the world before them, or whose presence belongs to time immemorial, and that—through the active manifestations of that presence—shaped the conditions of life for those living since. The ancestors *are* their stories, and it is the role of descendants to carry these stories on. Likewise, as people and animals go through life together, they become part of one another's stories, and as lives overlap, so they are carried on, or relayed, from generation to generation. Kinship, in this understanding, is fashioned in the relaying of lives, rather than set down in advance, in the inheritance of attributes. It is, if you will, the work of memory, of bringing stories of past lives into present experience in forging the relations of the future. Kinship is the memory of others lodged deep within ourselves, and the memory of ourselves lodged in the others. Our kinship with the earth and its creatures, thus, has nothing to do with sensibilities that have come down to us as part of an evolved legacy. We won't find it as a genetic endowment, inherited from ancient hunter-gatherers. Nor is it necessarily tinged with positive affect: with nonhuman as with human kin, we can love them and loathe them, often at one and the same time. What we cannot do, however, is escape them. For kinship is not the product of an evolutionary process but the matrix of relations in the unfolding of which life itself evolves. As such, it is the very condition of our existence in a relational world.

Back at the seaside, people and their animals come and go. Years pass. Knots are tied, untied, and tied again. The unending tapestry of kinship ravels and unravels. And all the while the waves, without a care for the world or its inhabitants, keep breaking on the shore.

NOTES

1. For further discussion of knots and knotting in relation to human kinship, see Tim Ingold, *The Life of Lines* (Abingdon, UK: Routledge, 2015), 18–26.

MAKING SHEEP-KIN AT THE END OF THE WORLD

Anne Galloway

For the past six years, I have lived with sheep—as few as four and as many as twelve at a time—but no local would call me a sheep farmer despite my engaging in many of the same practices that they do. Our small, feral-breed flock is too eccentric to be useful, and what should sheep be if not useful? After all, it is true that their utility is unsurpassed. Since sheep were domesticated over eleven thousand years ago, they have reliably provided humankind with meat, milk, fiber, hides, and comfort. Over millennia, we have transformed the Earth and each other, building shared worlds and making shared homes. Only in the past few hundred years have sheep ceased to also serve as so many humans' everyday companions through life and death, and the consequences of these diminished relations are significant.[1]

At any given time, there are least 1.25 billion sheep on the planet, representing a thousand distinct breeds and distributed across a wide range of physical environments.[2] Sheep are heavily entangled with the places and cultures of Judeo-Christian-Islamic tradition, as well as the expansion of the British Empire and scientific research.[3] Some flocks are farmed on lands so vast that the sheep are mostly wild and hunted for their meat. Others are intensively farmed indoors, their udders grasped by humans and machines once or twice a day. Some flocks walk beside their shepherds, across changing landscapes in a shared search for food and shelter; others are tethered in pens their whole lives. Yet sheep, no matter their geographical differences, are intimately related to our species in rich and often mutually beneficial ways.[4] Because one-quarter of Earth's human population relies on subsistence farming and still lives closely with nonhuman animals, it is not difficult to claim that sheep continue to take good care of people around the world.

Domesticated animals offer unique insights into kinship, and farmed animals have an exceptional ability to connect us with other animals, plants, lands, waters, and skies. For most of human history people have lived intimately with the animals they keep, and many still do.[5] But I live where the mass production and consumption of sheep are part of the natural order of things, and talk of "making kin" with sheep requires a rather particular sense of humor. For the past ten years, I have studied wool, meat, and milk sheep and their kin in this small island nation of Aotearoa/New Zealand, making human and nonhuman friends along the way, experiencing things that gutted me and lifted me up. And I am fascinated by what it might mean to "be in good relation" with a creature so "easily dismissed" and "killable" that it usually counts only in terms of the weight of its disassembled parts.[6]

I brought my fieldwork home when I had the opportunity and means to "help conserve" some of the country's endemic, feral breeds. Eccentric sheep, generally kept by eccentric people for eccentric reasons, "feral sheep are now far more similar to each other than any of them are to modern commercial sheep in New Zealand."[7] But I chose hybrid vigor over purity, and our flock is mostly redomesticated Arapawa, with some redomesticated Pitt Island, and probably a little Merino as well. One of the scientists working to sequence the sheep genome kept a small, personal flock of Arapawa sheep, and her research found that their closest relatives are Gulf Coast Native sheep.[8] Like the Navajo Churro, they are descendants of Spanish pre-Merino flocks brought to the Americas in the 1500s, with their own multispecies political entanglements.[9] In the mid-twentieth century, Gulf Coast sheep provided most of the raw wool for the southern United States, but they were gradually replaced by larger, more "productive" breeds and their survival now relies on rare-breed conservation.[10]

Arapawa sheep suffer the same perceived lack: they are small and slow-growing animals, not well suited to mass production. The first shore-whaling station was set up on Arapaoa in 1827, and sometime after, one of many American boats left their home sheep on the island. Some went feral and became a genetically distinct group found nowhere else in the world. Their redomestication largely arose in response to a government cull to protect native wildlife. Pitt Island feral sheep descend from Australian Merino, but they are also a genetically distinct breed. More numerous than the feral Arapawa flock, they faced culling for competing over grass with the island's commercially farmed sheep. Redomesticated Pitt Island sheep remain notoriously wild, and so also not well-suited to mass production. A few years ago, a Welsh sheep farmer asked me if it was ethical to redomesticate an animal that had finally escaped us. I still don't have an answer for him, but living with sheep has changed my understanding of domestication, kinship, and what an embodied, lived ethos might look and feel like.

We are valley and hill dwellers in the south of Aotearoa/New Zealand's Te Ika-a-Māui North Island, feet and hooves on ancient mudstone and clay soil, surrounded by native bush pierced by introduced pasture grasses, watched over by Kapakapanui rising a kilometer above us. If you had been traveling by train in late 1932, you might have looked out the window at the hills in the distance and read out loud from the carriage copy of *NZ Railways Magazine*, "Kapakapanui is a forested peak regarded by the Maoris [*sic*] as a 'lightning mountain' of omen; lightning flashing downward repeatedly above or near its summit was regarded as a portent of a chief's death or other misfortune to the tribe living below."[11] Despite persistent rain, lightning strikes have been far and few between since we moved here. But I still think of this story every time it storms, and there are now no Māori living here to ask.

The valley is ancient but has known only human footsteps since the early days of Māori settlement in the thirteenth century. The first people walked this land listening for the moa, kererū, and tūi, and although the giant moas are long gone, we still hear the thunder of kererū flight and arresting tūi song. With its headwaters in the Tararua range, the Ngatiawa River flows past us on its way to the sea, and climbing the hills requires walking through cold, quick-flowing waters. We share our water with occasionally seen eels and crayfish and always-heard insects, amphibians, birds, and small mammals, both native and introduced.

Five hundred years after the first Polynesians called this land home, British missionaries and settlers displaced or destroyed most of the local inhabitants, and by the 1850s, enough land had been cleared of native forest for sheep to hit the ground running. By the turn of the century, the area became known for logging and sawmills, and most cleared land was converted to dairy farming. The local post office and school closed decades ago, and neither logging nor farming is the primary source of local income, but the valley is full with sheep again.

We are part of the latest wave of settlers to this land, living along a road of mostly professional, foreign-born landowners, on what New Zealanders call a "lifestyle block," or hobby farm. We mortgage the smallest of several properties radiating off a dead end, all backing onto hill-country conservation land that gives each of our homes a larger-than-life feel. There are commercial smallholders in the valley, but we produce only what we can consume, and we are all legally prohibited from selling homegrown sheep meat for food safety reasons, as homekills are not vetted by government inspectors. Bushwalkers now climb a Department of Conservation trail to a hut on Kapakapanui, and I look out across the valley to hills of native bush and cleared, green paddocks spotted with small white figures. The weather tends toward hot and dry in summer and wet and cold in winter, but these patterns are changing, and trackers are sent out every year to rescue overconfident humans from the land and water, forest and sky.

The vital energy of the valley is balanced with everyday death. Wild creatures visibly and invisibly come and go from our shared lifeworld. Human hunters stalk feral pigs, goats, and deer, and fur trappers regularly check their lines, all in the name of native species conservation and self-sufficiency. To protect native wildlife, as well as the introduced ducks from whom we gather eggs, we kill rats, stoats, and possums, and our cat grows fat on rabbits. We feed the blood, skin, and bone to our vegetable garden and the often-circling hawks. What the ewes do not eat of their placentas gets absorbed by the earth, alongside the occasional stillborn lamb and old sheep, feeding all the soil creatures. And like almost all the people around me, once a year I kill some of the sheep to make room for new life, to eat and feed others. It may seem counterintuitive to some, but being responsible for the lives and deaths of other creatures can strengthen the bonds of kinship, recalibrate unequal relations, and hold us accountable for not taking more than we give.

This kind of holistic nourishment and reciprocal care is at the heart of what I believe to be good relations among humans, sheep, and the earth to which we both belong. While nourishment and care are certainly part of New Zealand's sheep meat, milk, and wool industries, they tend to be recognized, measured, and rewarded only in material or economic terms. The kind of relation I am talking about evades easy measurement and is less amenable to policy and regulation, but this kinship can be lived and is already being lived.

In Māori cosmology, the world is not a resource for humans, but humans are part of the world and creation is a dynamic process, always becoming "networks of kinship and alliance" and "animated reciprocal exchanges."[12] To live in, and with, the world is to recognize that all things have a *tinana* (body), a *wairua* (spirit), a *mauri* (life force), and a *hau* (cosmic vitality). Together, they "protect *tapu* [the sacred] and so maintain the *mana* [authority] of the person or group, the tree or forest, the dandelion or flower, the stream or ocean."[13] Insufficient or unbalanced care of the life force of one affects the life forces of all, and death comes when the *mauri* can no longer hold the whole together. A Māori approach to sustainable farming is described as interconnected and interdependent:

> A positive relationship between humans and a river would be evidenced by human land management practices that enable a river to maintain and enhance its *mauri*, which would result in its life-generating capacities being maintained. In this way, the *mauri* of the river is grown or maintained through ensuring that its life-generating vibrancy is not diminished. Simultaneously, the *mauri* of people is maintained through the provision of food and other resources to the humans from the river. Through maintaining the *mauri* of the river, the *mana* or dignity of the river is seen to be maintained, while the *mana* or dignity of the people is maintained through being provided for by the river.[14]

The average Māori farm is four times the size of the average New Zealand farm, and half of Māori farms are grasslands supporting sheep and beef cattle. In 2018, "Māori farms with sheep had just over four times as many (6,822 on average) compared with all New Zealand farms with sheep."[15] Not only does large-scale sheep farming play an important role in Māori community and food sovereignty; Māori created the world's first Indigenous organic verification system. By acknowledging soil as a living entity, the Hua Parakore framework goes beyond traditional, "Western" organic requirements "to ground their practices in a Māori worldview, expressed through six main *Kaupapa* [principles]: *whakapapa* [genealogy], *wairua* [spirit], *mana* [authority], *māramatanga* [insight], *mauri* [life force] and *te ao tūroa* [the greater world] . . . If the *wairua* [spirit] of the land is good, that's because the *kaitiaki* [caregiver] is looking after it properly. And there is a roll-on effect to the *kaitiaki* and their *whanau* [family]."[16]

Almost all sheep farmers I have met articulate a duty of care to the animals and the land, but care takes many shapes.[17] Even at large scale, sheep in te ao Māori are held in fundamentally different relation than are the sheep I have encountered during my fieldwork on non-Māori farms. Although I'm inspired by the Māori system of caring for living *taonga,* or treasures, the kind of kinship and reciprocal care I experience with our flock is different again from either of these examples, and I wonder what kind of "response-abilities" I have to our flock if I consider its members kindred.[18]

Ursula is ten years old now, the matriarch of the flock, answering to both her name and Big Mama. Like me, she is past breeding age and gray at the edges, enjoys watching the world go by, and gets stiff after sitting too long. But earlier today, I watched her run around, jump high in the air, and kick her legs out, playing with her offspring and their offspring, and I knew she was joyful in her body and with her kin in the world, which is definitely something that sheep do well. At the beginning, Ursula protected her lambs by stomping and snorting and chasing off the neighbors' large dog, never to return, and I've seen that solitary fierceness in service of others every time she has led an offensive or assembled a defense since. She does not like being handled for shearing or hoof trimming or needles of any kind, but she endures them all. She always approaches me when I enter the paddock for a few treats from my pocket, just for her, and enjoys having her flank patted. Ursula is no longer the first onto fresh grass, but the rest refuse to settle until she is with them. We will all know when her life ends, and I will leave her body with the flock until they have had a chance to spend what moments they need to with her.

I look at, listen to, touch, and smell the same six sheep every day. I originally brought home Ursula and her triplet ewe lambs, Grace, Emmaline, and Mercy. All four have given birth multiple times on this land and have never been out of one another's lines of sight or sound. If I understand their social world correctly, Grace will be the next matriarch. She has always been my friend, strong and curious and affectionate from the first day. But making sheep-kin is complicated. I killed Grace's first and last lambs for food, and castrated the one in between so he could stay with the flock. Gus and Grace still stick together like glue three years later, and they are among the great loves of my life.

Emmaline is grumpy and boisterous and difficult and beautiful, as was her daughter Edith, as is her daughter Esther. They are the relations who amuse and frustrate me no end. As lambs they climbed trees, and now they need cattle-height pens to keep them, and all attempts at care are met with great resistance. But they are lively and interesting companions, and under the brightness and heat of the sun, Esther closes her eyes and falls lightly asleep under the touch of my hand—the greatest display of trust any sheep can give you. Emmaline is next in our breeding rotation, and I will most likely kill those lambs for food, too. After Emmaline rejected him, I bottle-fed Esther's twin, Ned, and we were the best of friends until he died suddenly and unexpectedly at eleven weeks. I sat beside his burial cairn yesterday, as I often do, thinking about how losing him led me to keep Gus and Esther, and wondering whether Ursula could be buried nearby.

Mercy has always been everyone else's favorite, the only white one with dark spots in a flock of black sheep, friendly, gentle, and pretty as a picture. And now she is alone with her mother, her sisters, and their lambs, simply because I killed all of hers but one, whom I failed to keep alive. Mercy has always enjoyed a bit of solitude, but she makes it clear she does not entirely trust me and that seems fair. Still, I will not take another lamb from her, and since she cannot know that, I protect her food from the other sheep who now push her around, and that grants me permission to touch her. Yesterday she stood still and let me scratch her all over, walking away when she had enough.

Making sheep-kin means knowing how to care for body and spirit, *tinana* and *wairua*. All the ewes and Gus like being scratched under the chin and having their ears rubbed, but I avoid their foreheads since I do not want to be headbutted or gored by horns. I always let them avoid or walk away from me when I approach, but if I stop petting Gus too soon, he taps me with a front hoof until I resume. I feel along their spines and pelvises, searching for fat losses or gains, although their lean breed does tend to carry fat in the belly instead. I look at their eyes to see if they are anemic, and in their mouths to see if their teeth are in good condition. The sheep are fully shorn once a year, before the summer heat, then vaccinated and crutched or trimmed during winter pregnancy to keep the vulva and udders clean and accessible for lambing season in early spring. The sheep and shearer have an openly hostile relationship but necessarily push through their encounters. Hoof trimming is a more regular task that I do myself, and even that cannot prevent occasional injury or foot-rot. I provide antibiotics only when infection is present, and every few years I need to treat some internal or external parasites on one or two of the sheep. Only twice have the sheep suffered fly-strike, four years apart but in the same family, and we caught Emmaline and Esther before the maggots broke the skin and started eating them alive. All the sheep let me know when something is wrong—I just have to listen.

Making sheep-kin means knowing and caring for the land and water and sky we share, so they too can know and care for us. Our ways of giving back to the earth are varied and incomplete, but we take the responsibility seriously. There have been no synthetic fertilizers or herbicides used since we moved here, and a variety of new grasses, clovers, and herbs have been planted to provide a more varied diet and support a more diverse paddock ecology. The sheep and the soil are regularly fed seaweed and mineral supplements to keep them healthy and resilient to the increasingly volatile environment. They have access to fresh water at all times but not to waterways. In a country where livestock shelter is considered a bit of an oxymoron, the sheep have tree shelter in each paddock, plus a built shelter in the main one, and they regularly use both. There is something very pleasing about entering the paddock and seeing six heads pop out of the open shed to say hello, and sitting in the shed with the lambs, watching a storm together, is the safest I have ever felt.

To be allowed to sit with sheep, to earn their trust and have them accept you as their kindred, requires modulating your own body and spirit. Sheep move slowly, steadily, and deliberately—until they startle and bolt. If you move like they do, and if you learn to "speak sheep," they will invite you to share their rich and peaceful lifeworld. If you do not, they will struggle against you. To speak sheep is really to look and listen, to know that their ears and feet communicate more than their eyes, and to always give them a way out and away from you. As I glance out the window, I can see Ursula sitting on a small hill under a tree. From that position she can see any threats and warn the flock. But they are all safe, and she quietly regurgitates her food. If I were beside her, I would hear her body rumble, pause, and gently burp, then go back to chewing. Sheep breath is hot and smells of bitter plants, very slightly fermented. Their bodies smell of lanolin, soil, and shit, sunshine and rain. I keep small bits of raw fleece to remind me that I am bound to these creatures and them to me on this small island at the end of the world—and there is nowhere I would rather be.

NOTES

Arohanui to Professor Joanna Kidman (Ngāti Maniapoto, Ngāti Raukawa) for sharing family stories about Arapawa sheep-kin and for listening to mine.

1. David Abram, *Becoming Animal: An Earthly Cosmology* (New York: Pantheon Books, 2010).
2. Food and Agricultural Organization of the United Nations (FAO), "Sheep," http://www.fao.org/livestock-systems/global-distributions/sheep/en/.
3. On tradition, see Michael Lawson Ryder, *Sheep & Man* (London: Duckworth, 1983). On the British Empire, see Rebecca Jane Houghton Woods, *The Herds Shot round the World: Native Breeds and the British Empire 1800–1900* (Chapel Hill: University of North Carolina Press, 2017). On scientific research, see Sarah Franklin, *Dolly Mixtures: The Remaking of Genealogy* (Durham, NC: Duke University Press, 2007).
4. Alan Butler, *Sheep: The Remarkable Story of the Humble Animal That Built the Modern World* (Hants, UK: O Books, 2006).
5. Radhika Govindrajan, *Animal Intimacies: Interspecies Relatedness in India's Central Himalayas* (Chicago: University of Chicago Press, 2018).

6. See Kim TallBear, "Caretaking Relations, Not American Dreaming," *Kalfou: A Journal of Comparative and Relational Ethnic Studies* 6, no. 1 (2019): 24–41; Philip Armstrong, *Sheep* (London: Reaktion Books, 2016); Vinciane Despret, *What Would Animals Say If We Asked the Right Questions?*, trans. Brett Buchanan (Minneapolis: University of Minnesota Press, 2016).
7. D. F. G. Orwin and A. H. Whitaker, "Feral Sheep (*Ovis aries* L.) of Arapawa Island, Marlborough Sounds, and a Comparison of Their Wool Characteristics with Those of Four Other Feral Flocks in New Zealand," *New Zealand Journal of Zoology* 11, no. 2 (1984): 201–24.
8. Natalie K. Pickering, E. A. Young, J. W. Kijas, D. R. Scobie, and J. C. McEwan, "Genetic Origin of Arapawa Sheep and Adaptation to a Feral Lifestyle," *Proceedings of the Association for the Advancement of Animal Breeding and Genetics* 20 (2013): 451–54, http://aaabg.org/aaabghome/AAABG20papers/pickering20451.pdf.
9. Marsha Weiseger, *Dreaming of Sheep in Navajo Country* (Seattle: University of Washington Press, 2011).
10. Wikipedia, "Gulf Coast Native Sheep," https://en.wikipedia.org/wiki/Gulf_Coast_Native_sheep (last accessed April 7, 2021).
11. J. Cowan, "Through Storyland—Along the Wellington-Manawatu Line—Scenes of Beauty and Tales of the Past," *New Zealand Railways Magazine* 7, no. 6 (1932): 25, http://nzetc.victoria.ac.nz/tm/scholarly/tei-Gov07_06Rail-t1-body-d9.html.
12. Manuka Henare, "*Tapu, Mana, Mauri, Hau, Wairua*: A Maori Philosophy of Vitalism and Cosmos," in *Indigenous Traditions and Ecology: The Interbeing of Cosmology and Community*, ed. John Grim (Cambridge, MA: Harvard University Press, 2001), 197–221, at 198.
13. Henare, 210–12.
14. John Reid, Tremane Barr, and Simon Lambert, "Indigenous Sustainability Indicators for Māori Farming and Fishing Enterprises: A Theoretical Framework," *NZ Sustainability Dashboard Research Report 13/06* (ARGOS, 2013), https://www.nzdashboard.org.nz.
15. Stats NZ, *Profits Almost Double for Māori Farming Businesses* (2020), https://www.stats.govt.nz/news/profits-almost-double-for-maori-farming-businesses.
16. Jessica Hutchings, Jo Smith, and Garth Harmsworth, "Elevating the Mana of Soil through the Hua Parakore Framework," *MAI Journal* 7, no. 1 (2018): 92–102; Rebecca Reider, "Māori Organic Kai," *NZ Herald*, November 17, 2015, https://www.nzherald.co.nz/element-magazine/news/article.cfm?c_id=1503340&objectid=11547402.
17. María Puig de la Bellacasa, *Matters of Care: Speculative Ethics in More Than Human Worlds* (Minneapolis: University of Minnesota Press, 2017).
18. Donna Haraway, *Staying with the Trouble: Making Kin in the Chthulucene* (Durham, NC: Duke University Press, 2018).

ZACUANPAPALOTLS

Brenda Cárdenas

In memory of José Antonio Burciaga, 1947–1996

We are space between—
the black-orange blur
of a million Monarchs
on their two-generation migration South to fir-crowned
Michoacán where tree trunks sprout feathers,
a forest of paper-thin wings.

Our Mexica cocooned
in the membranes de la Madre Tierra
say we are reborn zacuanpapalotls, mariposas negras y
anaranjadas
in whose sweep, the dead whisper.

We are between—
the flicker of a chameleon's tail
that turns his desert blue backbone
to jade or pink sand,
the snake skinned fraternal twins
of solstice and equinox.

The silvering dusk, ashen dawn,
la oracíon as it leaves the lips,
first moment of sleep,
the glide into dreams
that husk our mestizo memory.

We are—
one life passing through the prism
of all others, gathering color and song, cempazuchil and drum
to leave a rhythm scattered on the wind,
dust tinting the tips of fingers
as we slip into our new light.

PERMISSIONS

These credits are listed in the order in which the relevant contributions appear in the book.

"A Prayer to Talk to Animals" is used with the permission of Nickole Brown. © 2018 by Nickole Brown, *To Those Who Were Our First Gods* (Rattle, 2018).

Martin Lee Mueller's "These Wild, Vibrant, Unstoppable Expressions of Aliveness" includes "Without'" (seven-line excerpt) by Gary Snyder, from *Turtle Island*, copyright © 1974 by Gary Snyder. Reprinted with the permission of New Directions Publishing Corp.

Selected passages from Merlin Sheldrake's "Can We Grow the Concept of Our Selves?" were excerpted from *Entangled Life* by Merlin Sheldrake, published in the United States by Random House, an imprint and division of Penguin Random House LLC, copyright © 2020 by Merlin Sheldrake.

"Boy I" and "Boy II" were originally published in Heather Swan, *A Kinship with Ash* (West Caldwell, NJ: Terrapin Books, 2020).

"Zacuanpapalotls" was originally published in Brenda Cárdenas, *Boomerang* (Tempe, AZ: Bilingual Press, 2009) and is used here with permission.

ACKNOWLEDGMENTS

As editors, we want to offer some deeply felt gratitude. An initial gathering in 2018—enabled by the generosity of the Center for Humans and Nature—began the conversations that would eventually become the *Kinship* series. During our time together, our group of twenty or so people often sat in a loose circle—some in chairs, some leaning against couches, some cross-legged on the floor—listening to unique experiences from all varieties of places. About this gathering, mostly it is the joy we remember, the laughter that seemed to bubble up spontaneously, reinforcing bonds of kinship and love for this living earth as we attempted to put this into words and actions.

We also recall the ways other voices came into our midst. During one of the meetings, for example, a participant in the middle of the circle pulled out her iPhone and gently asked us to listen. The sound of a family of Orcas communicating with one another entered the room. Even those of us who don't live anywhere near the Pacific Ocean felt deep recognition, made more poignant by the current threats faced by these fellow mammals. They were speaking to one another, yet it felt as though their voices were also reaching through the water to us. Kin.

Later, we ventured outside to stretch our legs and to participate in a soundwalk to attune ourselves to the nearby forest and its aural textures. We paused at a huge Oak whose sprawling crown seemed to cover most of the backyard. In a way, this long-lived Oak was the reason we gathered where we had. This tree's presence preceded the home in which we were meeting by many decades, likely centuries, and the home was there because the parents of

Strachan Donnelley, the founder of the Center for Humans and Nature, chose to live in this spot. The family story has it that this elder Oak tree drew them to the place. We paid our respects. Some of us laid a hand on the tree. All of us breathed, thinking about plant kin who outlive us in age and who gift us with oxygen.

On the basis of this meeting, the initial vision for *Kinship: Belonging in a World of Relations* was the creation of a single volume. Then it grew. As each of us, as editors, reached out to people of various expertise, asking them to share their perspectives and stories of kinship, new threads were suggested to us and the web became larger. Soon it was apparent that the web could not be contained by a single volume—at least one that abided by the rules of standard publishing specs. We decided to ask, What if? What if we let form follow function? What if we let this book become what it wants to become? What if this book should be a series? This line of questioning begat logistical problems. For one, publishers in an already-risky landscape can't comfortably take those kinds of chances.

Yet challenges can also create innovations. The *Kinship* series has thus become the Center for Humans and Nature's first venture into book publishing—of many, we hope. We're thrilled with the outcome. We'd like to express our deep gratitude for human kin that helped bring this ambitious project to such a beautiful result. Emily Lonigro, Demetrio Cardona-Maguigad, Felix Castellanos, Carla Levy, and Stacey Saunders of LimeRed, for the gorgeous book covers and visual design. Minds blown. Manuscript editor Katherine Faydash, for her expert eye and enthusiastic support of the content. Riley Brady, for the lovely page layout and design. Ronald Mocerino at the Graphic Arts Studio Inc., for meeting our every printing need. Chelsea Green Publishers, especially Michael Weaver and Michael Metivier, for being our fine confidants in distribution and promotion. Paul and Sandy Quinn, for hosting us so graciously at Windblown Hill. For their giant spirits, supportive presence, and care-filled work, the Center for Humans and Nature kinfolk: James Ballowe, Hannah Burnett, Anja Claus, Katherine Kassouf Cummings, Jon Daniels, Brooke Hecht, Bruce Jennings, Curt Meine, and Jeremy Ohmes; and for the support of the Center for Humans and Nature board: Gerald Adelmann, Julia Antonatos, Jake Berlin, Ceara Donnelley, Tagen Donnelley, Kim Elliman, Christopher Getman, Charles Lane, Thomas Lovejoy, Ed Miller, George Ranney, Bryan Rowley, and Eleanor Sterling.

Gavin would like to give extra thanks and love to his family—Marcy, Hawkins, and Peanut—for indulging him as a "nature nerd" and for keeping his heart full. Also, Coyote, Magpie, and the Crab-like Orbweaver—the world wouldn't be the same without you.

Robin offers gratitude to her human and more-than-human kinfolk for their loving support: Family, Students, Maples, Orioles, Foxes, Peepers, and the whole dazzling web of relations.

John would like to thank his wife, Karen, for her near quarter century of partnership in cultivating kinship, from raising our daughters Atalaya and Sol to sharing work as teachers in the School of Environment and Sustainability at Western Colorado University to building Camp Alpenglow in the heart of the Gunnison Country. None of John's work is possible without the mountains and snows that frame and stand sentinel above his valley, providing greater-than-human family since he first visited from New Jersey at age sixteen.

CONTRIBUTORS & KIN, VOLUME 3

Dr. Sharon Blackie is an award-winning writer and internationally recognized teacher whose work sits at the interface of psychology, mythology, and ecology. Her highly acclaimed books, courses, lectures, and workshops are focused on the development of the mythic imagination and on the relevance of our native myths, fairy tales, and folk traditions to the personal, social, and environmental problems we face today. As well as writing four books of fiction and nonfiction, including the best-selling *If Women Rose Rooted,* her writing has appeared in the *Guardian,* the *Irish Times,* the *Scotsman,* and more, and she has been interviewed by the BBC and other major broadcasters on her areas of expertise.

Nickole Brown is the author of *Sister* and *Fanny Says.* She lives in Asheville, North Carolina, where she periodically volunteers at several different animal sanctuaries. Her work speaks in a queer, Southern-trash-talking way about nature beautiful, damaged, and in desperate need of saving. The first collection of these poems, *To Those Who Were Our First Gods,* won the 2018 Rattle Chapbook Prize, and in 2020, her essay-in-poems, *The Donkey Elegies,* was published. In 2021, Spruce Books of Penguin Random House published *Write It! 100 Poetry Prompts to Inspire,* a book she coauthored with her wife, Jessica Jacobs.

Brenda Cárdenas's books and chapbooks include *Boomerang* (Bilingual Press); *Bread of the Earth/The Last Colors,* with Roberto Harrison; *Achiote Seeds/Semillas de achiote,* with Cristina García, Emmy Pérez, and Gabriela Erandi Rico; and *From the Tongues of Brick and Stone* (Momotombo Press). She also coedited *Resist Much/Obey Little: Inaugural Poems to the Resistance* (Spuyten Duyvil Press) and *Between the Heart and the Land: Latina Poets in the Midwest* (MARCH/Abrazo Press). Cárdenas has served as Milwaukee's poet laureate, cotaught the inaugural workshop for Letras Latina's Pintura : Palabra, A Project in Ekphrasis, and is associate professor of English at the University of Wisconsin–Milwaukee.

Photo by Jeanne Theoharis

Ourania (Nia) Emmanouil is a first-generation settler-Australian with Macedonian ancestry who was born on the traditional lands of the Wurundjeri First Nation. She is an interdisciplinary researcher and educator (philosophy, Indigenous knowledges, education, law) who predominantly works in intercultural contexts with First Nations and settler peoples. Guided by the principles of relationality and respect, Nia's work explores ontological shifts that support emergent forms of "listening" to place and the potential to use such shifts to cocreate with people-place.

Monica Gagliano is a research associate professor in evolutionary ecology and former fellow of the Australian Research Council. She is currently based at Southern Cross University, where she directs the Biological Intelligence Lab funded by the Templeton World Charity Foundation. She has pioneered the brand-new research field of plant bioacoustics, for the first time experimentally demonstrating that plants emit their own "voices" and detect and respond to the sounds of their environments. Her work has extended the concept of cognition (including perception, learning processes, memory) in plants. Her latest book is *Thus Spoke the Plant* (North Atlantic Books, 2018).

Dr. Anne Galloway is a multispecies anthropologist, speculative design ethnographer, founder of the More-Than-Human Lab (www.morethanhumanlab.nz), and associate professor in design for social innovation at Victoria University of Wellington, New Zealand. Anne's cultural research focuses on human-livestock relations, and she designs the conditions that make new relations between humans and animals possible. You can find her on Twitter @annegalloway or with her loved ones, both human and nonhuman.

John Hausdoerffer, jhausdoerffer.com, is author of *Catlin's Lament: Indians, Manifest Destiny, and the Ethics of Nature,* as well as coauthor and coeditor of *Wildness: Relations of People and Place* and *What Kind of Ancestor Do You Want to Be?* John is dean of the School of Environment & Sustainability at Western Colorado University and cofounder of Coldharbour Institute, the Center for Mountain Transitions, and the Resilience Studies Consortium. John serves as a fellow and senior scholar for the Center for Humans and Nature.

Photo by Keith Carlsen Photography

Sean Hill is the author of *Dangerous Goods* (Milkweed Editions, 2014) and *Blood Ties & Brown Liquor* (UGA Press, 2008). He has received numerous awards, including fellowships from the Cave Canem Foundation, Stanford University, and the National Endowment for the Arts. Hill's poems and essays have appeared in several journals and in nearly two dozen anthologies, including *Black Nature* and *Villanelles*. He directs the Minnesota Northwoods Writers Conference at Bemidji State University. Hill lives in Montana with his family and is a visiting professor of creative writing at the University of Montana. More information can be found at his website: www.seanhillpoetry.com.

Photo by Geoff Wyatt

Julian Hoffman is the author of *Irreplaceable: The Fight to Save our Wild Places*, which was the Highly Commended Finalist for the Wainwright Prize for Writing on Global Conservation in 2020. His first book, *The Small Heart of Things*, was selected by Terry Tempest Williams as the winner of the 2012 AWP Award and went on to win a National Outdoor Book Award for Natural History Literature. He lives beside the Prespa lakes in northern Greece.

Photo by Jon Webber

Tim Ingold is professor emeritus of social anthropology at the University of Aberdeen. He has carried out fieldwork among Saami and Finnish people in Lapland, and has written on environment, technology, and social organization in the circumpolar North, on animals in human society, and on human ecology and evolutionary theory. His more recent work explores environmental perception and skilled practice. His books include *The Perception of the Environment* (2000), *Lines* (2007), *Being Alive* (2011), *Making* (2013), *The Life of Lines* (2015), *Anthropology and/as Education* (2018), *Anthropology: Why It Matters* (2018), and *Correspondences* (2020).

Dr. Robin Kimmerer is a mother, botanist, writer, and Distinguished Teaching Professor at the SUNY College of Environmental Science and Forestry in Syracuse, New York, and the founding director of the Center for Native Peoples and the Environment. She is an enrolled member of the Citizen Potawatomi Nation and a student of the plant nations. Her writings include *Gathering Moss* and *Braiding Sweetgrass: Indigenous Wisdom, Scientific Knowledge and the Teachings of Plants*. As a writer and a scientist, her interests include not only restoration of ecological communities but also restoration of our relationships to land. She lives on an old farm in upstate New York, tending gardens domestic and wild.

Christopher (Toby) McLeod has been project director of the Sacred Land Film Project since 1984. In 2013, he completed the four-part series *Standing on Sacred Ground*, which aired on PBS. Toby produced and directed *In the Light of Reverence* (2001) and three other award-winning documentaries broadcast on national television: *The Four Corners: A National Sacrifice Area?* (1983), *Downwind/Downstream* (1988), and *NOVA: Poison in the Rockies* (1990). His first film was *The Cracking of Glen Canyon Damn*—with Edward Abbey and Earth First! McLeod has a master's degree in journalism from University of California, Berkeley. Toby has worked with Indigenous communities as a filmmaker, journalist, and photographer for forty-four years.

Martin Lee Mueller is a postdoctoral researcher at the Department of Teacher Education and School Research at the University of Oslo. His Nautilus Book Award–winning debut, *Being Salmon, Being Human,* inspired the stage performance *Discovering the Wild in Us and Us in the Wild,* which has been touring internationally since 2016. Martin is cofounder of the nonprofits Small Earth Institute and VILLAKS, and affiliate scholar with the Arne Naess Programme at the Oslo-based Centre for Development and the Environment. Martin's writing interweaves moral philosophy, Earth systems science, Indigenous perspectives, ecoliteracy, and ecopoetry.

Photo by Birgitta Eva Hollander - Heia Folk

Steve Paulson is the executive producer of the Peabody Award–winning public radio program *To the Best of Our Knowledge.* His radio reports have also been broadcast on NPR's *Morning Edition* and *All Things Considered.* He has moderated an annual series of public events at the New York Academy of Sciences, including "The Emerging Science of Consciousness," "Beyond the Big Bang," and "The Power of Wonder." Steve has written for *Salon, Slate, Nautilus, Chronicle of Higher Education,* and *Los Angeles Review of Books.* His book, *Atoms and Eden: Conversations on Religion and Science,* was published by Oxford University Press.

Richard Powers is the author of a dozen novels. His most recent, *The Overstory,* won the 2018 Pulitzer Prize for Fiction.

Merlin Sheldrake is a biologist and author of *Entangled Life: How Fungi Make Our Worlds, Change Our Minds, and Shape Our Futures.* He received a PhD in tropical ecology from Cambridge University for his work on underground fungal networks in tropical forests in Panama, where he was a predoctoral research fellow of the Smithsonian Tropical Research Institute. Merlin is a keen brewer and fermenter, and is fascinated by the relationships that arise between humans and more-than-human organisms. Find out more at merlinsheldrake.com, or follow him on Instagram (@merlin.sheldrake) or Twitter (@MerlinSheldrake).

Photo by Hanna-Katrina Jedrosz

Dr. Eleanor Sterling is Jaffe Chief Conservation Scientist at the American Museum of Natural History's Center for Biodiversity and Conservation. Her current work focuses on the intersection of biodiversity, culture, and languages and the development of indicators of well-being in biocultural landscapes. She is a world authority on aye-ayes, nocturnal lemurs found only in Madagascar. She has a lifelong fascination with languages and has studied over ten languages, from Vietnamese to Swahili. She is an adjunct professor at Columbia University's Department of Ecology, Evolution, and Environmental Biology.

Heather Swan is the author of the poetry collection *A Kinship with Ash* (Terrapin Books) and the nonfiction book *Where Honeybees Thrive: Stories from the Field* (Penn State Press), winner of the Sigurd F. Olson Nature Writing Award. Her nonfiction has appeared in many publications, including *Aeon, Belt, Catapult,* and *Minding Nature.*

Her poetry has appeared in *Terrain, Poet Lore, The Hopper, Phoebe, Cold Mountain Review, Midwestern Gothic, Wildness,* and several anthologies. She has been the recipient of the Martha Meyer Renk Fellowship in Poetry, the August Derleth Prize, and an Illinois Art Council Fellowship. She teaches writing and environmental literature at University of Wisconsin–Madison.

Gavin Van Horn is the creative director and executive editor for the Center for Humans and Nature. His writing is tangled up in the ongoing conversation between humans, our nonhuman kin, and the animate landscape. He is the co-editor (with John Hausdoerffer) of *Wildness: Relations of People and Place,* and (with Dave Aftandilian) *City Creatures: Animal Encounters in the Chicago Wilderness,* and the author of *The Way of Coyote: Shared Journeys in the Urban Wilds.*

Manon Voice is a native of Indianapolis, Indiana, and is a multi-hyphenate—poet and writer, spoken-word artist and film producer, hip-hop emcee, educator, social justice advocate, community builder, and practicing contemplative. She has performed on diverse stages across the country in the power of word and song and has taught and facilitated art and poetry workshops widely. Her poetry has appeared in the *Flying Island, Indianapolis Review,* the *House Life Project: People + Property Series, Sidepiece Magazine, The World We Live(d) In* anthology, *The Indianapolis Anthology,* and more. In 2018, Manon received a nomination for the Pushcart Prize in Poetry. Manon is also a 2021 Editorial Fellow with the Center for Humans and Nature.

Rowen White is a seed keeper and farmer from the Mohawk community of Akwesasne and a passionate activist for indigenous seed and food sovereignty. She is the educational director and lead mentor of Sierra Seeds, an innovative Indigenous seed bank and land-based educational organization located in Nevada City, California. Rowen is the founder of the Indigenous Seedkeepers Network. She believes through the power of cultivating creative supportive learning spaces, reclaiming narratives, and practicing radical imagination, we can work together to seed the change for a more equitable and beautiful, relational, kincentric food system that centers around a deep sense of belonging and connection.